Spaces of Colonialism

RGS-IBG Book Series

The *Royal Geographical Society (with the Institute of British Geographers) Book Series* provides a forum for scholarly monographs and edited collections of academic papers at the leading edge of research in human and physical geography. The volumes are intended to make significant contributions to the field in which they lie, and to be written in a manner accessible to the wider community of academic geographers. Some volumes will disseminate current geographical research reported at conferences or sessions convened by Research Groups of the Society. Some will be edited or authored by scholars from beyond the UK. All are designed to have an international readership and to both reflect and stimulate the best current research within geography.

The books will stand out in terms of:

- the quality of research
- their contribution to their research field
- their likelihood to stimulate other research
- being scholarly but accessible.

For series guides go to www.blackwellpublishing.com/pdf/rgsibg.pdf

Spaces of Colonialism

Delhi's Urban Governmentalities

Stephen Legg

Blackwell
Publishing

BLACKWELL PUBLISHING
350 Main Street, Malden, MA 02148-5020, USA
9600 Garsington Road, Oxford OX4 2DQ, UK
550 Swanston Street, Carlton, Victoria 3053, Australia

The right of Stephen Legg to be identified as the Author of this Work has been asserted in accordance with the UK Copyright, Designs, and Patents Act 1988.

First published 2007 by Blackwell Publishing Ltd

1   2007

Library of Congress Cataloging-in-Publication Data

Legg, Stephen.
Spaces of colonialism : Delhi's urban governmentalities/
Stephen Legg.
p. cm. — (RGS-IBG book series)
Includes bibliographical references and index.
ISBN 978-1-4051-5633-2 (hardcover : alk. paper) —
ISBN 978-1-4051-5632-5 (pbk. : alk. paper)
1. Delhi (India)—Politics and government—20th century, 2. New Delhi
(India)—Politics and government. 3. Delhi (India)—Historical
geography. 4. New Delhi (India)—Historical geography. I. Title.

DS486.D3L37 2007
320.954'5609041—dc22

                                          2006033545

A catalogue record for this title is available from the British Library.

Set in 10/12 Plantin
by Newgen Imaging Systems (P) Ltd, Chennai, India

The publisher's policy is to use permanent paper from mills that operate a sustainable forestry policy, and which has been manufactured from pulp processed using acid-free and elementary chlorine-free practices. Furthermore, the publisher ensures that the text paper and cover board used have met acceptable environmental accreditation standards.

For further information on
Blackwell Publishing, visit our website:
www.blackwellpublishing.com

Contents

Figures

Tables

Abbreviations

BCE	Before the Common Era
CRS	Communal Riot Scheme
DAGS	Delhi–Ajmeri Gate Scheme
DIT	Delhi Improvement Trust
DMC	Delhi Municipal Committee
DTPC	Delhi Town Planning Committee
ICS	Indian Civil Service
IDC	Imperial Delhi Committee
ISA	Imperial Secretariat Association
NDDC	New Delhi Development Committee
NDMC	New Delhi Municipal Committee
PWD	Public Works Department

Archival References

CC	Chief Commissioner's Files
DA	Delhi State Archives, New Delhi
DC	Deputy Commissioner's Files
EUR/MSS	European Manuscripts Collection
FR	Fortnightly Reports on the Political Situation in Delhi Province
IORL	India Office Records and Library at the British Library, London
IPI	Indian Political Intelligence files (edited by A.J. Farrington, microfiche published by IDC, London, 2000)
NA	Indian National Archives, New Delhi
NMML	Nehru Memorial Museum and Library, New Delhi
SASL	Centre of South Asian Studies Library, Cambridge

Series Editors' Preface

Like its fellow RGS-IBG publications, *Area*, the *Geographical Journal* and *Transactions*, the series only publishes work of the highest quality from across the broad disciplinary spectrum of geography. It publishes distinctive new developments in human and physical geography, with a strong emphasis on theoretically informed and empirically strong texts. Reflecting the vibrant and diverse theoretical and empirical agendas that characterise the contemporary discipline, contributions inform, challenge and stimulate the reader. Overall, the book series seeks to promote scholarly publications that leave an intellectual mark and that change the way readers think about particular issues, methods or theories.

Kevin Ward
University of Manchester, UK

Joanna Bullard
Loughborough University, UK
RGS-IBG Book Series Editors

Preface

My attention was first drawn to Delhi by the brash arrogance yet undeniable beauty of the new capital city, constructed between 1911 and 1931. As I became more aware of the post-colonial literature regarding the colonial archive and at times, the unwitting complicity of present intellectual formations with those of the past, I grew uneasy regarding the representations of New Delhi. Portrayals of the city communicated little of the teeming life, confusion and buzz that are the hallmarks of the urban condition, even such a fabricated and dispersed condition as that created in the capital. More importantly, there was a marked absence of evidence regarding the resistance that always accompanies such unabashed demonstrations of authoritarian power. It became apparent during my visits to Delhi, in 1997, 2001 and 2003, that not only was the story of New Delhi incomplete, but that such a story would also have to incorporate the influence and history of the pre-existing walled city to the north.

As such, this manuscript traces three paths through the colonial spaces of Delhi. The first orients a geographical route between the monumental and residential spaces of New and Old Delhi. This path treks from the capital into New Delhi's residential and policed zones and hierarchies, explores the military and cordon sanitaire between the two cities and terminates in the spaces of surveillance and improvement in Old Delhi.

This movement necessitates the negotiation of a second, historiographical journey. This pathway leads from the literature of architectural and town planning history to the broader writings regarding colonial urbanism, and from there to the sprawling plains of post-colonial theory. The vastness of the latter terrain and its tilt towards literary theory have justified only tentative engagements with this work, although the influence of post-colonialism can be felt throughout. While inflecting the nuance and much of the detail of this work, the complexity and range of data that was collected regarding the government of New and Old Delhi could not be competently analysed using post-colonial theory alone. Rather, a system of analysis more persistently orientated towards practice, which could simultaneously incorporate the importance of materiality, knowledge and power, was required.

The writings of Michel Foucault provide a structuring analytic with which to negotiate the path from New to Old Delhi. His work on the 'governmentalisation of the state' combined an emphasis on the manoeuvrings of traditional political figures with a focus on the technologies, techniques, rationalities and knowledge formations that constitute the state itself. However, the relationship of Foucault's later works to his earlier writings is a complex one, as is the suggested relationship between modern biopower and the forms of power relation that pre-dated it. Such complexities necessitate a third path to be trodden through Foucault's intellectual biography, selecting a toolkit with which to piece together Delhi's spaces of colonialism. This path leads from his earlier, archaeological, work on classification and discourse to his later, genealogical, work on discipline and government. It also forces a mapping of Foucault's travelling theory, charting his applicability to the colonial world. Yet, this route is also the least linear of the three paths. Drawing especially on Foucault's lecture course of 1978, the co-constitution of sovereign, disciplinary and governmental forms of power is stressed, as are the temporal continuities between his various writings. Special thanks must go to Graham Burchell who kindly allowed me to consult a draft of his translation of Foucault's *Security, Territory, Population* lecture course. The theoretical interconnections made in this translation helped me portray the structured complexity of life in Delhi and also helped provide a geographically oriented guide to the empirical applicability of Foucault's writings.

These three paths should prove interesting to a range of scholars, from those looking for a regional historical geography of the capital of Britain's most-prized colony, to those interested in (post)colonial urbanism, or those seeking further evidence of the value of Foucault, and governmentality studies specifically, to critical, historical geographical analysis. Each of the three main chapters combines empirical analysis of a particular governmental landscape of ordering with an integrated discussion of the Foucauldian power relation that informed that ordering. Chapter 1 sets the scene of the imperial capital, discussing the heightened importance of sovereign power in Delhi and stressing the compatibility of such power with colonial governmentalities. Chapters 2–4 examine the residential landscape of New Delhi using Foucault's archaeological methodology, the policing of the two cities in terms of disciplinary diagrams and the biopolitical improvement of Old Delhi. The partial unity and coherence of these landscapes and the functioning of Delhi as a node in national and international networks are commented upon in Chapter 5 (Conclusion).

This research has drawn upon the support of innumerable people and institutions, not all of whom can be given the credit they deserve here. The majority of the work was carried out at the Department of Geography in the University of Cambridge, during a doctorate at Fitzwilliam College and a Junior Research Fellowship at Homerton College. The ESRC funded my doctorate and also provided a 1-year Postdoctoral Research Fellowship. Extra support for research and travel came from the Smuts Fund and the Allen, Meek and Reed Award at Cambridge, the Dudley Stamp Memorial Fund from the Royal Society and the Overseas Conference Fund of the British Academy. In Delhi, I enjoyed the support of Dr Sachidanand Sinha at the Jawaharlal Nehru University and the Sarai Initiative at the Centre for the Study of Developing Societies.

Special thanks must go to the staff who aided this research at the following locations; in the United Kingdom, the University Library, Centre of South Asian Studies and Department of Geography at Cambridge and the Oriental and India Office Collection at the British Library; in Delhi, the National Archives, the Nehru Memorial Museum and Library and, especially, the Delhi State Archives. Phillip Stickler was tirelessly patient in composing the excellent diagrams for Chapters 3 and 4; thanks also go to Chris Lewis for the extra help with diagrams throughout.

The three members of my doctoral supervisory committee have immensely influenced this research through their joint interest in colonialism and post-structural theory, and their friendship. Thanks go to Jim Duncan, for his guidance on post-colonial and landscape theory; Phil Howell, for his persistent Foucauldianism; and Gerry Kearns, for his inspirational commitment to political critique and urban historical geography. Alison Blunt and Mike Heffernan proved to be great mentors since examining my thesis, and thanks go to Stephen Daniels, David Matless, Alex Vasudevan and Charles Watkins for their support since my move to Nottingham.

This work has been discussed with various colleagues at conferences and symposia over the years. Thanks go to John Agnew, Chris Bayly, Rachel Berger, Stuart Corbridge, Nancy Duncan, Stuart Elden, Matt Gandy, David Gilbert, Derek Gregory, Narayani Gupta, Matthew Hannah (special thanks for his detailed reading of Chapter 2), Philippa Levine, Rajiv Narain, Anthony King, Satish Kumar, Colin McFarlane, Miles Ogborn, Si Reid-Henry, Richard Smith, Karen Till and Hannah Weston.

Beyond these formal circuits of debate are the endless social and academic networks of support that have been vital to sustaining this project. At the very least, these networks have included Sara AlNajjar, Sumit Baudh, Andy Bradburn, Ed and Rosanna Cox, Mark Day, Sacha Deshmukh, Naomi Dobraszczyc, Florent Le Flamanc, Millie Glennon, Hayley Milne, Nick Megoran, Lucy Norton, Tom Nutt, Kate Pretty, Neil Sinclair, Aunty and Uncle Singh, Mitu Sengupta and Robin Daniel Whittaker.

Finally, this manuscript could not have been completed without the help of Mum, Dad and Emma. While their trials and tribulations during these (many) years have more than outstripped my own, their support in sickness and in health for this absentee son/brother has been unwavering. Nice one.

Chapter One

Imperial Delhi

New Delhi was one of Britain's most spectacular showcases of imperial modernity. It was commissioned in 1911 to facilitate the transfer of the capital of British India from Calcutta to Delhi and took 20 years to construct. It embodied the rationality of imperialism in its aesthetics (refined, functional classicism), science (a healthy, ordered landscape) and politics (an authoritarian, hierarchical society). As a node within a global, imperial network of sights, New Delhi represented Britain's vision for an empire of legitimacy and longevity in the twentieth century.

The material reality of these utopic visions, however, did not prove acquiescent to imperial will. At the level of administration, bureaucracy and governance, Delhi's colonial landscape was as much dominated by the older city to the north of the imperial headquarters. This was *Shahjahanabad*, the walled city that had functioned as the capital of the Mughal Empire from 1648 to 1857. As against the neo-classical monumentalism of the imperial capital, and the sterile, geometric spaces of New Delhi, 'Old Delhi' was depicted as an organic space of tradition and community. Urban life here was conducted in congested and winding streets between communities defined by historic location and caste. Temporal flows were dictated by calls to prayer and a thriving annual schedule of Hindu, Sikh, Jain and Muslim festivals. Bereft of extensive modern sanitation and infrastructure, Old Delhi was a haptic and sensory place of smells, sights and contact that bewildered and beguiled Western tourists and governors alike.

This, at least, is the popular conception of the colonial geography of Delhi; of dual cities. This is embodied in the now iconic aerial photo of the dividing line between the two cities (see cover image). This book will explore the extent to which the two cities were, in fact, governed as one and impacted upon each other in myriad ways. As a closer inspection of the aerial photo shows, the *cordon sanitaire* between the two cities was, in fact, traversed by multiple, well-worn tracks. Similarly, streets within the old city had been widened and cleared, whereas the plot of land within the walled city to the west had been demolished in the nineteenth century and was reconstructed in the 1920–30s. These spatial traces hint at the geographies of interaction and incursion between the two cities.

Rather than plotting an entire history of Delhi as the capital of the Raj (1911–47), three case studies will be used to explore the interactions between the cities. These will show that, in terms of residential accommodation, policing and infrastructural improvement, the two cities were intimately intertwined. While being very different projects, these landscapes of interconnection shared similar political rationalities of practice that must be explored. Likewise, each landscape presents evidence of a colonial government that sought security and profit for itself over the welfare and development of the Indian population, and thus demands some sort of critical commentary.

The writings of Michel Foucault provide a toolkit with which to explore the complementarity of these seemingly diverse practices of rule. The governmentalities that infused spaces of residence, policing and improvement allow an analysis that maintains their specificity but suggests continuities in the thought, vision, identity politics, technology and ethos that informed them. Secondly, the body of literature that has sought to extend Foucault's writings to the colonial context suggests a number of ways in which colonial governmentalities can be articulated to critique imperial rule at the level of the everyday and the material. Having outlined these two bodies of literature, we will return at the end of the chapter to Delhi to set the historiographical and historical–geographical context for this empirical exploration of Foucault's later works.

Security, Territory, Population

Governmental rationalities

Underwriting the majority of Foucault's works are his ruminations on the concept of power, which became more explicit in his later, genealogical writings. Whereas the earlier, archaeological works had always been about power to an extent (see Chapter 2), the genealogical works of Foucault's later career addressed power directly and with a distinct terminology, in relation to material, governmental, social and spatial formations. Foucault referred to 'domination' as a structure of force in which the subordinate have little, or no, space for manoeuvre (Hindess, 1996: 97). In opposition, 'power' referred to a structure of actions that bears on the decisions of free individuals, making power unstable and reversible. Between these two forces lie the relations of 'government', the conduct of conduct that aims to regulate the behaviour of individuals and populations. Such power–knowledge relations are integrated in particular institutions, from the state to the family or a system of morality. As such, Deleuze (1988: 89) suggests that in each historical – and geographical – formation, we must ask what belongs to which institution, what power relations are integrated, what relations occur between institutions and how these divisions change from one stratum to the next (over time and space).

It is this fragmented and shifting vision of power that recurs throughout Foucault's thought, if not in the writings about him or much of his earlier published material. He identified modern forms of power that constituted, circulated and normalised without the central coordination of an ultimate sovereign. As such, there is not just 'power'; there are types of power relation that depend upon the forces, knowledges, archives and diagrams they relate for their characteristics. Yet, Foucault did refer to the types of power that emerged in the modern era as 'biopower'; powers over life that targeted both the individual body, through techniques of discipline, and the social body, through government of the population (Foucault, 1979b). This governmental regulation was exerted through various domains that were posited as autonomous to the state. These included the economy, society and the population, that last of which was targeted through 'biopolitics'.

These power relations cannot be neatly separated. Rather than successions or substitutions of power relations, there are changes of mode, moods and moments (Dillon, 2004: 41). Foucault rejected interpretations of his early work that stressed temporal discontinuity: he emphasised the difficulty of clean breaks (Foucault, 1970: 50, 1972: 175, 1980: 111), suggested that different rationalities 'dovetailed' together (Foucault, 1975–6 [2003]: 242) and later suggested that different forms of power entered into a form of triangulation, but that 'government' power attained a pre-eminence over sovereign and discipline power (Foucault, 1978 [2001]: 220).

These forms of power also retain complex relationships with their outsides, seemingly excluding subjects from the political order, only to include them more completely in politics by their outcast state. Such relations include the figure cast beyond the protection of the sovereign's law, the abnormal excluded from society through enclosure within disciplinary institutions or the uncivilised subject deemed incapable of liberal conduct.

Each chapter in this book will examine a particular landscape that was forged primarily through the forces of one particular type of power relation: the hierarchies of knowledge in New Delhi, disciplinary power and policing, and the biopolitics of urban improvement. These types of power relation will be addressed in detail in each chapter. Yet, throughout these forms of power, the persistent effects of sovereign power were felt. To foreshadow its consideration in the following chapters, sovereign power in the context of biopower will now be explained, in advance of a discussion of the translations of biopower to the colonial context.

Sovereign power

Sovereignty is an intensely territorial concept. From an original association with pre-modern empires, it came to refer to the post-Westphalian (1648) system of states within which there was one absolute authority who could legitimately exercise violence (Taylor, 2000, although see Elden, 2005 for a discussion of the complexity of the Treaty of Westphalia). International diplomatic relations determined that no state would intervene in the domestic politics of another without invitation, as the basis of mutual recognition of sovereignty. Within a territory, sovereignty could be exercised by the monarch or succeeding bureaucracies. Yet these superior powers, depended upon the consent of their subjects, which they offered up in return for certain rights and protections (Hindess, 1996: 12).

Foucault (1975–6 [2003]: 23–42) argued that this predominant institutional role of the sovereign had cast a juridical shadow over considerations of power relations in the post-medieval period. Juridical power attempts to prevent a type of action through the threat of legal or social sanctions and, as such, was still pitched as a concept that could be owned or possessed by the head of a hierarchy of rights and consent (Tadros, 1998: 78). This disregarded the new disciplinary mechanisms of power that had emerged (see Chapter 3 for a discussion of sovereignty, law and discipline). Foucault (1975–6 [2003]: 241) posed the sovereign as the body that only exercised its power over life when it extracted people, resources and taxes, or made a decision about killing; it had the right to take life or let live. Foucault encouraged us to look for power beyond the centre, beyond the realm of conscious decisions, as something that circulates and is not owned, and to begin our analyses with infinitesimal mechanisms, material operations and forms of subjection (Foucault, 1975–6 [2003]: 34). This would reflect the evolution of power relations towards an intrusive and self-formative biopower; a power over life itself.

Foucault's suggestions have been read by many as a call to abandon analyses of sovereign power in favour of endlessly circulating, anonymous forms of normalisation. Yet others have shown how the paradox of sovereignty continues to play itself out within the framework of contemporary biopower (Connolly, 2004). The paradox refers to societies in which the rule of law is enabled and secured by a sovereign that is above the law itself. Drawing upon yet challenging Foucault, Agamben (1998) has insisted that the sovereign has *always* been concerned with biopolitics, and that sovereign power retains the right to decide on a state of exception. Thereby individuals or groups are proclaimed to be beyond the protection of the sovereign's laws, and are thus exposed, as 'bare life', to violence without protection. Agamben has done a huge amount to reinsert considerations of violence and sovereignty into theoretical debates, yet his suggestion that exceptionalism and the (concentration) camp mark the *nomos* (the principles governing human conduct) of modernity surely presents an over-simplified and nihilistic approach to power relations. We can counter this simplification by continuing to address sovereign power in terms of resistance, complexity, its geographies and its varied imbrications with biopower.

First, an exceptionalist view of sovereign power provides little consideration of the possibility of *resistance*. Foucault (1975–6 [2003]) suggested that such resistance could occur at the level of counter-discourses that challenge views of society predicated on sovereign understandings of power, stressing society as a place of continuing war and bare life, not just peace and political life (see Neal, 2004). At a more embodied level, Edkins and Pin-Fat (2004) have suggested that resistance to sovereign power would target the attempt to divide life and the following production of bare life. The refusal of distinctions would challenge the act of counting and classifying, yet resisting bare life would mean accepting this status in an attempt to highlight the violent operation of sovereign power, as mobilised in non-violent non-cooperation. However, these considerations of resistance are constrained by an overly prescriptive understanding of sovereignty that reduces it to the power of exception.

Agamben empties sovereignty of much of the *complexity* of its practice and principles and reinstates a central model of power, over-emphasising the decision of the sovereign at the cost of the multiplicity of force relations operating in society (Neal, 2004: 375). The sovereign idiom of power conceals itself within capillaries of power and knowledge production. While sovereignty exposes itself in violence and terror, it can also be productive and generous in multiple, provisional and always contested ways (Hansen, 2005: 172). Starting with the writings of Jean Bodin from 1576, de Benoist (1999) has charted the variety of different forms of sovereignty. From an original basis in the ability to legislate, these forms have evolved through absolutist, revolutionary, nationalist, liberal and totalitarian regimes. Hansen and Stepputat (2005: 7) also used Bodin to sketch the non-exceptionalist characteristics of sovereign power that, besides the rights of law and war making, included office appointment, fiscal validation, taxation, language and land rights. In his book entitled *State of Exception*, Agamben (2005: 23) sought to stress the complex topographies of these exceptional spaces, but he operates within the definitions of the juridical order, not the actions these orders initiate in material or social space.

Sovereignty is a result of these actions, an ontological effect made real by ritualistic and performative evocations of power. Sovereign rights have been democratised such that citizens can now effectively wield them, although this also works to reinforce the adjudicator of these rights, which is often the sovereign power itself. Although still dependent on the ultimate ability of the sovereign to wield violence, this creates a much more fragile view of state sovereignty:

> sovereignty of the state is an aspiration that seeks to create itself in the face of internally fragmented, unevenly distributed and unpredictable configurations of political authority that exercise more or less legitimate violence in a territory. (Hansen and Stepputat, 2005: 3)

This more complex and fragmented view of sovereignty forces a discussion of its *geography* that is foreclosed by Agamben's insistence that the essence of sovereignty is the decision regarding exceptions. Walker (2004) has argued that this ignores the time–space specificities of sovereignty, reproducing Schmitt's absolute spatialities in which exceptions are fixed on passive space. Rather, sovereignty is spatiotemporally specific in its practices and complex sites (see Gregory, 2007). The topology of sovereignty is, thus, not a space, but a dividing practice that seeks to impose authority, the law, and often violence (Dillon, 2004: 56).

Hansen's and Stepputat's logical progression from their interpretation of sovereignty was to seek out its historical specificity in particular terri- tories. This involves studying historically embedded practices and cultural meanings of sovereign practice and violence, whether the latter is actual or borne in rumours and myth. Yet, to grasp these complex sovereignties means to fathom them in their articulation with modern forms of biopower. Although Foucault did argue forcefully for moving conceptions of power away from the sovereign, in his writings on discipline and sovereign power, he stressed their coming together (Foucault, 1975–6 [2003]: 39). As Dillon (2004: 45) has suggested, sovereignty co-evolves around the 'terrains of existence' of biopolitics and discipline, crafting itself around different grids of intelligibility.

The *imbrications* of sovereign- and bio-powers are gaining increasing attention. Hansen and Stepputat (2005: 9) have examined how the democ- ratisation of sovereign rights and the creation of national citizenries were accompanied by the emergence of intensive and caring forms of 'welfare' cameralism that formed one of the earliest arts of government. Dillon (1995, 2004) has long insisted that governmentality and sovereignty are not oppositional but complementary, relying upon each other and feeding their power–knowledge needs. While the norms of government and the excep- tions of sovereignty are often juxtaposed, they actually depend upon and reinforce each other (Hussain, 2003: 20).

Yet, it would be a mistake to cast sovereignty as the villain of the piece against biopower's heroic stance of making live and letting die (Foucault, 1975–6 [2003]: 241). Dean (2002a) has drawn attention to what

Foucault (1979b) depicted as the 'dark side of biopolitics'. Though sovereign power kills, it also 'lets live', and though biopower 'makes live', it can also disallow life, introducing killing machines at the level of the population and making massacres seem vital (Foucault, 1975–6 [2003]: 254, 1979b). While the sovereign right to kill has found itself increasingly restrained, biopolitics has increased its remit to manage life:

> not by returning to the old law of killing, but on the contrary in the name of race, precious space, conditions of life and the survival of a population that believes itself to be better than its enemy, which it now treats not as the juridical enemy of the old sovereign but as a toxic or infectious agent, a sort of 'biological danger'.   (Deleuze, 1988: 92)

The publication of Foucault's 1978 lecture course on *Security, Territory, Population* (Foucault, 1978b [2007]) will do much to set the context of his already published 'Governmentality' (Foucault, 1978a [2001]) lecture and to further complicate and imbricate the triangle of sovereign, disciplinary and governmental power (for a discussion of the following year's lectures on economic liberalism, see Lemke, 2001). Here Foucault denies again that there is a clear transition from legal (sovereign) to disciplinary and then security (governmental) ages, but stresses that the techniques of the legal and disciplinary world were taken up by security mechanisms that seek to regulate populations (Foucault, 1978b [2007], 11 January). He also insisted on complicating the spaces associated with each form of power. Sovereignty did not just refer to empty territory, it concerned itself with the same multiplicity of people targeted by discipline and mechanisms of security. As such, in discussing the town plans that best represented the three forms of power, Foucault referred to Le Maître's *La Métropolitée* of 1682 (also see Rabinow, 1982). This unbuilt, city plan organised different social groups in relation to each other and placed the capital city in a geometrically central position in the national territory. It was to be an ornament, displaying the best a territory could offer, and as such has many parallels to the utopian elements of New Delhi (see Chapter 2). Yet, the Indian capital also imbricated other types of power relation, creating a complex landscape of sovereignty, government and discipline.

Discipline

For his discussion of discipline, Foucault (1978b [2007], 11 January) turned to the seventeenth century new town of Richelieu which focused more on the distribution of individuals than social groups. The town not only had elements of symmetry, but also included dissymmetry, to allow smaller quarters to spatially express social status. Unlike the capitalisation of territory under sovereignty, here the question was of structuring space. Foucault (1978b [2007], 18 January) later stressed that while sovereign power forbade and prohibited, proscribing the city and displaying its strength, discipline focused on what one must do, rather than the forbidden, imposing an order from within. This transition was discussed most dramatically in *Discipline and Punish* (Foucault, 1977) showing how sovereign violence was replaced with institutionalised supervision for the criminal. Spatial divisions, time tables, bodily regularisation and different forms of supervision were used to reform the inmates of these institutions. These techniques also swarmed through an increasingly disciplinary society in the hope of creating economically efficient yet politically docile subjects. Foucault did not seek out ideal types to represent disciplinary power. Rather, he traced out the generalities between different techniques that were used to respond to local objectives. This was in an attempt to detect the functions of disciplinary power, which he termed diagrams and most famously examined through the Panopticon (Deleuze, 1988: 72).

Disciplinary power will be discussed and empirically investigated in Chapter 3, the purpose here is to stress its links to other types of power relation. Despite discipline's much-advertised departure from the power relations of sovereignty, this does not make the two incompatible. Foucault (1975–6 [2003]: 260) showed how the racist state, Nazi Germany in particular, brought the classic mechanism of death into perfect coincidence with the discipline and regulation of biopower. Two years later, in a controversial departure from the more dramatic ruptures between sovereign and disciplinary power suggested in *Discipline and Punish*, Foucault (1978b [2007], 25 January) suggested that '... the panopticon is the oldest dream of the oldest sovereign'. As such it was both modern and archaic because the figure at the centre of the Panopticon exerted his, her, or its sovereignty over all the individuals in the machine of power: 'The central point of the panopticon still functions, in a way, as a perfect sovereign' (Foucault, 1978b [2007], 25 January). Sovereign and disciplinary powers could also be bridged by the state that took up mechanisms of discipline and used them in conjunction with the objectives of sovereign power (Foucault, 1977: 213).

Discipline and government were also explicitly linked as they both arose in response to the failure of sovereign mechanisms to deal with the consequences of industrialisation and demographic explosions in early modern Europe (Foucault, 1975–6 [2003]: 249–50). Discipline has an ambiguous relationship to government and security, being one part of the binary of biopower, yet also serving as an opposing pole to the regulation of free populations. Disciplinary mechanisms had been comparatively easy to establish from the seventeenth century onwards, focusing as they did on deviant bodies that were viewed as a threat to social order. Yet, the seething multiplicity of society still needed regulating, although without the intense economic and investment, and political intrusiveness, of disciplinary surveillance. The response was a series of regulatory mechanisms that sought to normalise society such that it would function efficiently and productively.

Although both discipline and regulation were initially termed acts of 'normalization', they were later distinguished (Foucault, 1978b [2007], 25 January). Discipline analysed individuals, places, times and actions in order to compare them to a pre-existing norm, to which they would then be trained to conform; what Foucault termed 'normation'. In contrast, regulatory mechanisms would examine cases, risks, danger and crises in society in order to calculate the probable norm at which society should function, and to which the unfavourable were brought in line, referred to as 'normalization'. The scale at which regulation operated (the whole population) necessitated the freedom of its subjects. While discipline is centripetal, the population mechanisms of security are centrifugal; while discipline seeks to regulate everything, security observes society and decides what is desirable (Foucault, 1978b [2007], 18 January).

While at the functional level, disciplinary and regulatory mechanisms remain distinct, at the level of technology and practice they intersect, as with discipline and sovereign mechanisms. Foucault (1978b [2007], 8 February) reaffirmed his earlier analysis of the transition from the isolated technique of the Panopticon to the generalised mechanism of panopticism (Foucault, 1977: 213). The displacement of attention outside of disciplinary institutions, where their function became external and their objects of knowledge became more general, raised the question of whether the techniques and strategies of discipline merely fell under the totalising institution of the state. It was in order to go beyond the state, as he had gone beyond disciplinary institutions, that Foucault turned to the study of mechanisms of security, the regulation of populations, and governmental rationalities (governmentalities).

Governmentalities

Foucault (1978b [2007], 8 February) described the governmentality project as seeking the general technology of power that assured the state's mutations, development and functioning. Governmental rationalities emerged that had political economy as their main form of knowledge, the population as their target for regulation and apparatuses of security as their essential mechanisms (Foucault, 1978a [2001]: 219). An apparatus, or *dispositif*, is a concrete assemblage of diverse elements with a particular purpose, specific targets, and controlling strategies (see Rabinow and Rose, 2003). Examining them involves cutting across distinctions of thought, practice and materiality, and studying them at the surface level of the everyday. Such studies focus on networks of tactics and strategies, not on some structurally hidden level of causation.

There is now an extensive literature regarding governmentality, and the intention here is not to recap this corpus (Burchell et al., 1991; Barry et al., 1996; Dean, 1999; Rose, 1999; Hannah, 2000; Joyce, 2003). The aim, rather, is to place the one previously published lecture (Foucault, 1978a [2001]), which directly addresses governmentality, in the context of the 13 other lectures that were entitled 'security, territory, population' but which Foucault suggested should have been called a 'history of "governmentality"' (Foucault, 1978b [2007], 1 February). The analytical categories emerging from the governmentality literature that will structure this book will then be described, ahead of an evaluation of the failure of the governmentality literature to sufficiently address the significance of place, resistance, internationalism and criticism.

'Governmentality' refers to three things (Foucault, 1978a [2001]: 219–20, 1978b [2007], 1 February):

(1) *Power.* The emergence and pre-eminence, over discipline or sovereignty, of government as a type of power, which led to certain apparatuses and knowledges.
(2) *Analytics.* The ensemble formed by institutions, analyses, calculations and tactics that allow population to be targeted through political–economic knowledge and apparatuses of security.
(3) *Governmentalisation of the state.* The transition from the medieval state of justice to the administrative state.

The governmentality lecture (the fourth in the series, Foucault, 1978b [2007]) emphasised written works from the sixteenth century regarding the 'arts of government'. Yet, the opening three lectures actually introduced the governmentality concept through the practical measures that emerged in response to a changing political, demographic and geographical reality, not through the mentalities or abstract rationalities of government. This placed the emphasis on the security apparatuses that served to regulate the free movement and circulation of the objects of government. The early lectures addressed, first, spaces of security in relation to town planning; second, the 'event' and uncertainty through a discussion of grain trade and human morbidity and, finally, normation/normalisation with regard to vaccination.

The 'governmentality' lecture itself then traced the movement from earlier arts of government, which still advised the sovereign and sought to control and supervise the population, to political sciences that sought to observe and regulate the population from a distance. This denoted a shift from the government of things, organising their *disposition* so as to lead them to an end that profited the sovereign, to a government of *processes* that had their own end and internal logics. The emergence of the economy and the population as concepts, with corresponding political–economic and biopolitical realities that were independent of the sovereign, marked the transition from arts to sciences of government (Curtis, 2002; Legg, 2005a). Biopolitics has been somewhat fetishised in the literature, overshadowing the regulation of other domains of government such as society or the economy (Dean, 2002a: 48). Such domains were obviously interconnected, but the rationalities that were devised to govern them often came into conflict whereby, for instance, free market economics threatened to cause social disruption, or biopolitical schemes to regulate the population proved too expensive.

One of the most formative conflicts was that associated with the rise of liberalism, which Foucault examines as an active art of government rather than a political philosophy (Gordon, 1991; Rose, 1996). Liberalism facilitated the democratisation of rights not only against the sovereign, but also against overly intrusive disciplinary acts of surveillance or 'over-government'. Security apparatuses sought to protect the 'liberty' of free subjects so as to defend supposedly natural economic, social or demographic processes. Yet, these apparatuses simultaneously allowed the acquisition of knowledge about those they sought to protect from over-government, allowing them to normalise any non-self-regulating individuals. As the urban environment displayed ever more pungent and distressing signs of the failure of liberalism in late nineteenth- and early twentieth-century Europe, attempts were made to integrate individuals into a 'society' that would be subject to state programming but distant from it.

The rest of the lectures took in the genealogy of pastoral power, the diplomatic–military technique and that of the 'police' of the seventeenth to the eighteenth centuries. The history of the pastorate was examined over five lectures, tracing how this power that individualised and cared for all emerged from the Hebrew–Christian tradition. In attempting to target the intimate level of conduct, pastoral power faced resistance and forms of counter-conduct (Foucault, 1978b [2007], 1 March). Yet, through the Reformation and Counter-Reformation, pastoral government was taken up into the emergent 'state reason' of the seventeenth century.

While Foucault devoted much time to his genealogy of the governmentalisation of the state, it is the resultant ensemble of practices and analyses that allow the population to be regulated that are of immediate interest here. Governmentalities are not just 'govern mentalities'; they also refer to the operationalisation of knowledge, technologies of representation and the execution of a political imaginary (Dillon, 1995: 333). As such, they should be examined not only through discourse analysis but also through more thoroughgoing analytics. Rabinow (1982: 269) defined this as the isolation of historical characteristics that permit us to see how a grid of intelligibility enables actions to proceed. Such an analysis admits that governmental projects need not have an all encompassing and unifying rationale, but that consistencies within limits recur across different forms of government. An analytic of governmentality stresses that government predominates, but does not decimate, previous types of power relation, and thus that the categories of analytic investigation can be applied to both the poles of biopower and sovereign power. The works of Rose (1996), Dean and Hindess (1998) and Dean (1999) identify dimensions of analysis that enable an investigation of specific manifestations of a governmentality:

(1) *Episteme*. Distinctive ways of thinking and questioning; the use of certain vocabularies and procedures for the production of truth.
   (a)   Which forms of thought, calculation or rationality are deployed?
   (b)   How does thought seek to transform practices?
   (c)   How do practices of governing give rise to specific forms of truth?
(2) *Identities*. The epistemological conception of the people to be governed, their statuses and capacities, the shaping of agency and direction of desire.
   (a)   What forms of conduct are expected?
   (b)   What duties or rights do people have?
(3) *Visibility*. Ways of seeing and representing reality; practical knowledge of specialists and policy makers; plans, maps, diagrams.
   (a)   How are some objects highlighted whereas others are obfuscated?
   (b)   What relations are suggested between subjects and space?
   (c)   How is risk mapped and what are the suggested remedies?
(4) *Techne*. Techniques and technologies of government; ways of intervening in reality through strategies and procedures in relation to the materials and forces to hand and the resistances or oppositions encountered.
   (a)   Through which mechanism, procedure or, tactic is a rule accomplished?
   (b)   How are local contingencies incorporated and exploited?
(5) *Ethos*. The moral form that distributes tasks in relation to ideals or principles of government; the orientation invested in practices.
   (a)   Who benefits from a regime of government?
   (b)   Where and with whom are values invested?

A stable correlation across these dimensions suggests a taken-for-granted regime of practices that can be problematised and placed under a programme of reviews. Dean (1998: 185) explicitly suggests that problematisations should be central to an analytical approach that '... proceeds from an analysis of, if not their congenitally failing character, their local and particular instances of problematization and reproblematization'. Similarly, Rabinow and Rose (2003) suggest that apparatuses are initially formed in response to crises, problems or perceived challenges to those who govern, the resolution of which can lead to more generally applicable governmental rationalities. These problematisations embed the concept of resistance within a critical governmentality perspective, reinforcing a conception of power that draws upon points, knots or focuses of resistance to make change possible, linking it to the outside: '... the final word on power is that *resistance comes first*' (Deleuze, 1988: 89; emphasis in the original).

Just as apparatuses cross-cut divides between the material, performative and the cognitive, so do problematisations target the practical conditions that make something an object of knowledge (Deacon, 2000). But problematisation can also take into account the conditions of emergence of an apparatus and the technologies of self by which humans engage with it.

Genealogy itself is a form of problematising the taken for granted, and it feeds on historical, practical incidents of problematisation to string together its historical analyses. This was most explicitly highlighted in *The Use of Pleasure* (Foucault, 1986a), where problematisations of being and the practices they problematised were used to structure the moral investigation of 'sexual' pleasures in ancient Greece and Rome. Foucault (1986a: 36) suggested that sexual conduct was stylised in four ways (dietics of the body, economics of marriage, erotics of boys, and the philosophy of truth) but that these stylisations had certain fields of problematisation in common (questioning ethical substance, types of subjection, forms of elaboration of the self and moral teleology). Exploring this matrix allowed Foucault to examine one surface of emergence of Greek and Roman culture in the ancient world. Although attempts to replicate this analysis can risk becoming too overtly structuralist (Philo, 2005: 331, n. 2), it does allow the framing of empirical investigations that chart the operation of apparatuses which function along the lines of force that constitute different power relations. Such a matrix structures the chapters of this book. Each chapter, to borrow from Rabinow (1989: 14–15) marks an irruptive event that led to shifts in apparatuses of power relations (for comparable approaches, see Chatterjee, 1995, and Ogborn, 1998). Given the range of power relations, and the empirical detail of the case studies, it would be impossible to do justice to the full network involved in each apparatus of control. Rather, each chapter examines a particular form of landscaping as a spatial surface of emergence for each apparatus (see Table 1.1). The analytical categories of the governmentality literature will be used to structure investigations that seek to encompass the range of levels, from categories of thought to performative identities or material technologies, and the relations of power through which such apparatuses operate. These categories will not only be evident throughout the book but will also be returned to in the conclusion to draw out the analytical continuities within Delhi's governmental regime.

Table 1.1   Analytical chapter structure

Analytic	Residential	Policing	Improvement
Power relation	Classification	Discipline	Biopolitics
Episteme	Imperial urbanism	Colonial policing	Colonial urbanism
Identity	Coloniser	Criminal	Slum dweller
Visibility	Town plan	Riot scheme	Intensity map
Techne	Housing	Surveillance	Dispersion
Ethos	Hierarchy	Partition	Levelling

The selection of landscapes is necessarily arbitrary to an extent, and the lack of focus on trade, craftswork or industrial landscaping risks reproducing a shift away from economic analysis associated with Foucauldian studies more generally. However, Delhi did not have a large industrial labour force until the mid-1930s and its reputation as a centre for craftsmanship traded heavily on its past glories.[1] As such, the examined landscapes focus more on the efforts to secure the two cities made by the central and local governments. These had to respond to problematisations in the realm of accommodation, policing and urban heath that overshadowed overly 'economic' problems, although tensions with the rationalities of economy and finance recur throughout.

Limits of governmentality

Place?

Foucault had an innate interest in spatial relations (see Driver, 1985; Philo, 1992; Osborne and Rose, 1999; Elden, 2001). This interest expressed itself in the geometric language with which he dissected the archive, his interest in disciplinary spaces of incarceration and segregation and his studies of the use of space to regulate populations (see each chapter for discussions of archaeological, disciplinary and biopolitical approaches to space in Foucault's work). Yet, there have also been criticisms of his spatial formations. These have suggested that his archaeological, discursive works deploy spatial language while being divorced from the material realm, that his disciplinary diagrams rely too heavily upon plans and not upon constructed realities, and that the governmentality work fails to engage with the territories it claims to order.

For instance, Dupont and Pearce (2001: 133–5) accuse Foucault of 'objective idealism' for failing to appreciate the blocks and obstacles to the development of governmental rationalities, and of 'subjective idealism' for focusing too much on individual authors as opposed to non-subjective rationalities. These criticisms, in part, are due to not only an over-reliance on the one 'governmentality' lecture, but does also hint at a deeper-seated concern with Foucault's approach. This is that while he has an acute awareness of the geometry of power and the striations of social space, he underplays the messy aliveness of place. This is less in evidence in his historical studies than in his more abstract theorisation. As Rabinow (1982: 269) commented, Foucault used space as one of several tools to analyse power–knowledge relations, not always to study the space itself. Yet, spatial studies can open up complex realities in their focus on regimes of practices. Indeed, Dean (1998: 185) associates these regimes with studies of places where rules seek to guide what is done and said.

Foucault's spatial work, especially with regard to disciplinary institutions, is well known (see Chapter 3). The broader scale of governmental regulations means that their geographies are necessarily more diffuse and complex. Hannah (1997) has not only suggested how geographies of discipline and government may interact, but he has also provided a thoroughgoing analysis of governmentalisation of U.S.A. territory in the nineteenth century (Hannah, 2000). Processes of abstraction, assortment and centralisation helped gather the data that allowed the population and territory of the United States to be conceived of, and thus normalised. This normalisation has to take place at the local level, the attendance to which has led to calls for a 'spatial governmentality' (Merry, 2001) or 'realist governmentality studies' (Stenson, 2005). Yet, Macleod and Durrheim (2002: 43) have stressed that Foucault's work on governmentality did not see him abandon his previous commitment to an 'ascending analysis of power' that begins at the micro-scale. Rather, governmentality seeks to unite the local and the national, a type of government that should breed a form of analysis that pays attention to both micro, individualising, and subjectivising processes as well as those that both totalise and objectivise.

However, this matrix of power relations can create a sterile and lifeless depiction of place. Thrift (2007) has argued that Foucault's approach neglected a consideration of not only affect and inanimate 'things', but also of space itself. Order was prioritised over aliveness; the co-incidence, energy and motion of the world were demoted beneath its diagrams and grids of intelligibility. There is great force to these arguments, encouraging us as they do to think not just of Foucault's geometries of power, but also of his substantive, historical geographies (Philo, 1992).

There are a series of works that encourage us to do this. Huxley (2007) has shown that governmentalities have spatial and environmental logics that have casual effects, which operate through different modes (such as geometric ordering, environmental causality or social disposition). Drawing on untranslated material, Elden (2007) has shown how Foucault collaborated on a series of projects that examined urban infrastructures, hospitals and the politics of habitat in the mid-1970s as his thoughts on discipline and government were forming. An emphasis on towns as machines, metaphors, territories and spaces of political economy foreshadowed his later work not only on discipline, but also on security.

In his 1976 lecture on discipline and biopolitics, the emergence of the latter was linked explicitly to the 'urban problem' and the 'milieu' in which people lived (Foucault, 1975–6 [2003]: 245). This link was returned to 2 years later. It was stated emphatically that the problem of the town was at the heart of the different mechanisms of security (Foucault, 1978b [2007], 25 January), and that 'urban objects' were the essential condition for the rise of 'policing' as an art of government (Foucault, 1978b [2007], 5 April). In discussing the 'spaces of security', Foucault (1978b [2007], 11 January) contrasted the sovereign capitalisation of territory and the disciplinary structuring of space to the security-inducing planning of the milieu of events. A milieu, whether natural or artificial (i.e. of physical or human geography), was defined as a space that supported action through mass effects on the population who inhabit it. Circular links are created between effects and causes, and these processes are targeted by urban regulatory interventions. As such, the milieu represents a rare, explicit *spatiality* in Foucault's work, examining the co-constitutive relation of social and spatial relations.

Yet, where Foucault risks certain collusions with those he seeks to examine are in his operation within the limits of consideration set out by the governmentalities he studies. A detached approach to town planning gives little sense of how these plans were operationalised. An approach to grain and dearth focused around governmental policy gives us little sense of the pain of starvation or the panic buying of the urban market that grain doctrines induced (Foucault, 1978b [2007], 18 January). An emphasis on the emergence of the concept of a population that could only be affected by calculations and distanced controls removed attention from the very real forms of regulation that urban populations were at times exposed to (Foucault, 1978b [2007], 25 January).

In short, Foucault's emphasis on the spaces of power–knowledge provide fascinating insights into the regimes of practices that are being constructed, but fail to capture the complexity of the places over which these regimes are, not always successfully, extended. Places problematise the operation of apparatuses; they have a tendency to bleed, become infected, break, leak, collapse and also to foster innovation and conspiratorial spaces of counter-conduct or outright resistance. Places are the *excess* of space. But this resistance need not be conscious or human. Power can be exerted over space and, seeing as all power creates resistance, the intransigent landscape can be considered as a resistant and inseparable element of governmental apparatuses (Joyce, 2003: 185–6). It is this notion of space as a stubborn, alive and problematising medium that will inform the chapters the follow. Here, space is not only a medium of order, but one that also presented the climate and over-crowding of a badly misplaced New Delhi, the unmapped old city, a dangerously undocile nationalism, and the disease and congestion of the walled city.

Resistance?

The call for an appreciation of place in all its complexity does not mean that one must abandon the governmental archives in search of a counter, subaltern archive. Counter-discourses and, in this case, forces of anti-colonial nationalism were essential to problematising governmentalities and feature prominently in the archive. These movements can be traced exactly *by* their impact on the spaces of ordering that mark the surfaces of different apparatuses. This does not necessarily defuse them of their power or influence; such points of resistance can be analysed to highlight the extent to which seemingly omnicompetent apparatuses are vulnerable to attack, self-doubt and internal rupture. O'Malley (1996: 323) was right to warn against assuming that all resistance is internalised and neutered. The following chapters will show how programmes of government can fail, not only due to the stubborn materiality of place, but also due to the conscious resistance of those who did not stand to benefit from the colonial ethos.

Foucault has been criticised as the 'scribe of power' (Said in Said et al., 1993 [2004]: 214); explicating a process in which humanity becomes ever more panoptic, carceral, regulated, normalised, conducted and subjectivised. Hindess (1997: 261) has suggested that the problem begins with Foucault's use of the term 'political' and its conflation with the phrase 'governmental'. As opposed to an association simply with the government of the state, other traditions associate the political with the appropriation, redistribution or allocation of the powers of government. Given that modern power depended upon free subjects, individuals must retain the ability to resist. This 'plebeian aspect' springs from the variety of influences open to people as they craft themselves, or the tendency for error and miscalculation to create different, and critical, perspectives on taken-for-granted practices (Ransom, 1997: 117).

Although resistance was more explicitly addressed in Foucault's later work, a more consistent yet implicit emphasis on resistance has been detected by Pickett (1996). In his works on 'insanity', Foucault traced contestations and transgressions of societal limits. In the early 1970s, he wrote about local struggles and resistance as he turned to address power relations following the student French uprisings of 1968. This led to a more considered 'politics of resistance' in his works from the late 1970s onwards. Here, resistance was stressed as being essential to the practice of power relations and as something that arose from local, material conditions of existence:

> For, if it is true that at the heart of power relations and as a permanent condition of their existence there is an insubordination and a certain essential obstinacy on the part of the principles of freedom, then there is no relationship of power without the means of escape or possible flight. Every power relationship implies, at least in potential, a strategy of struggle, in which the two forces are not superimposed, do not lose their specific nature, or do not finally become confused. Each constitutes for the other a kind of permanent limit, a point of possible reversal.   (Foucault, 1982 [2001]: 346)

The horizontal unity of resistance was challenged by the segregation of disciplinary practices (Foucault, 1977: 219), but Foucault did elaborate on how individual resistance could be interlinked. Collective resistances had evolved from targeting domination in the feudal period to exploitation in the nineteenth century and subjection in the contemporary period (Foucault, 1982 [2001]: 331). Conceptions of resistance as such have to move on from the model of the revolution that would assume the position of sovereign power and should seek to string together local points of resistance: '... producing cleavages in society that shift about, fracturing unities and effecting regroupings, furrowing across individuals themselves, cutting them up and remoulding them, marking off irreducible regions in them, in their bodies and minds...' (Foucault, 1979b: 96). The most local of these forms of resistance takes place within the self, in which forms of counter-conduct emerge that challenge a regime's advocated government of the self (Foucault, 1978b [2007], 1 March).

Internationalism and criticism?

Foucault was undeniably Eurocentric. The pertinent question is whether this mattered, given that he did not make any claims to be a world historian or a transcendental philosopher? Those who apply Foucault outside of Europe surely bear the burden for translating his work themselves? While this is the case, European knowledges, technologies and epistemologies were dependent upon relations with their colonial 'outsides' (Legg, 2007). Indeed, to fail to appreciate this is to uncritically accept many of liberalism's most misleading arguments. For instance, Adam Smith insisted that value lay in bodies and not land, fostering dreams of perpetual growth. Yet, this dream was dependent upon the conception of the rest of the world as free space that could be appropriated at will (Cooper, 2004: 521). As such, the apparatuses of governmental security associated with liberal economics in

Europe historically relied upon a colonised zone of exception and the exertion of sovereign power over colonial territories. Similarly, Stoler (1995) has dismissed European bourgeois claims to racial superiority and an innately more advanced civilisation by showing how Europe itself was dependent on imperial circuits of identity formation in terms of race, gender and sexuality.

Such arguments will be addressed in the following section, but they necessarily inter-mesh with the debate about 'critical governmentality studies'. Foucault's politics are notoriously amorphous because he consistently refused traditional trajectories of investigation and imperatives to answer traditional political questions. Yet, his political relevance has also been dampened by studies that refuse to take him at his word when he insisted upon studying local struggles over power relations in all their materiality. O'Malley et al. (1997) describe this as an over-emphasis on governmentalities and a lack of emphasis on the 'messy actuality' of rule. This has dulled the critical edge of genealogy and risks turning governmentality studies into a defence of liberalism and a means for its renewal (Stenson, 1998). In so doing, one perpetuates liberalism's greatest achievement and its object; the conduct of conduct from a distance, not here over others through space or class, but over us through time and the archive.

The response is to examine governmentalities in all the detailed confusion of their places of elaboration and in the context of resistance that could both be internalised in programmatic reviews or remain resiliently external and hostile. Analytical categories can, as such, be used to provide not only a 'thick description' of government, but also a critical analysis, revealing disjunctures between governmentalities and practices within a complex topography of rule (Dean, 2002c: 120). This would highlight the continued presence of non-liberal power relations of sovereignty, discipline and biopolitics through an active and focused criticism.

Scott (1999) has faced the challenge of mounting a post-colonial criticism in the face of anti-foundational, post-structuralist suggestions that such criticism must rest on universalisation or must inscribe a new rationalism. His response is to pose criticism as a strategic practice in a contingent problem space that generates specific objects and questions. Each chapter of this book investigates the problem space of a different type of power relation and poses the problematisations that arose within the detailed context of place and resistance that threw the colonial government into question. In line with Dean and Henmann (2004: 492–3), the aim is to address the question 'where of power' by analysing the imbrication of different types of power relation within the zones, spaces and locales in which territories, their inhabitants and their products have been appropriated across the Earth.

Colonial Governmentality

In his influential historical introduction to post-colonialism, Young (2001: 4) insisted that post-colonial analysis should not privilege the colonial, but should examine the effects of colonialism in the present. This Foucauldian history of the post-colonial present would be oriented towards social justice and the contestation of domination. Yet, Young's work itself focuses on the very historical origins of post-colonial theory and, as such, sheds great light on current cultural politics and geopolitics. In a similar vein, while Scott (1999: 16) is committed to tracing the effects of colonial history in the present, he admits that this sort of analysis must begin with the rationalities of rule that colonialism established. This would move beyond a study of behaviour to one of social reforms that altered the terrain of struggle against governmental power. It would also move beyond an examination of resistance to look at how colonial power relations affected the terrain on which resistance could operate. The basis of a post-colonial critique can, thus, be an interrogation of the practices, modalities and projects through which the lives of the colonised were altered. This suggestion drew upon Scott's (1995, reprinted in Scott, 1999) earlier work on 'colonial governmentality' that tailored Foucault's work to the colonial world while retaining an emphasis on the practicalities of rule. This was a contribution to a growing field of literature of which only a rudimentary sketch can be presented here, although individual texts will be engaged with throughout the empirical investigations that follow.

The spatial differences of colonial governmentality

One approach to the difference of colonialism is to sketch a seemingly oxymoronic geography of liberalism. Despite its universal claims, liberty was only granted to those who were sufficiently normalised in line with regularities that emerged in the West, and were most readily apparent in adult, white, heterosexual, able-bodied men (Valverde, 1996; Mehta, 1999; Hindess, 2001; Dean, 2002b). Yet, the governmental rationalities of colonial rule require a more local and material form of analysis. Scott (1995: 192) introduced the approach of colonial governmentality as a means of moving post-colonial analysis beyond considerations of textual representations of authority, the denial of voice and the institutional mechanisms of colonial control. While not denying the importance of such analyses of how the colonised were included or excluded, the emphasis was shifted onto the ways in which colonial power was organised as an activity designed to produce effects of rule. Scott stressed the contingent nature and discontinuous developments within colonial government itself, as manifested in particular 'projects' (Thomas, 1994: 105).

At colonial governments' 'points of application' (Scott, 1995: 199), the emphasis lay on eradicating superstition and prejudice in spaces where socialisation occurred, and on erecting new conditions in their place laid out on clear, rational principles. These 'cultivated settings', in Helliwell's and Hindess's (2002) terms, aimed to produce governing effects on conduct, that is, to induce a milieu that would improve those who interacted in it: '*the systematic redefinition and transformation of the terrain on which the life of the colonized was lived*' (Scott, 1995: 205; emphasis in the original). Although when compared with the colonial '*commandement*' (Mbembe, 2001) of most African states, India appeared liberal, this was very much a translated and exported brand of liberalism. Studying this translation and disaggregation of state forms is essential if the explicitly 'Western' theories of liberalism and governmentality are to be critically and accurately studied outside the West (Hansen and Stepputat, 2002: 10). There has been a protracted debate, however, regarding what the 'difference' of colonial government actually is. The perspectives taken on this difference can, very generally, be navigated using the triangular powers of governmentality (see Table 1.2).

As Cooper (2004) has suggested, colonial governments operated in a more intimate relationship with the violence of *sovereign* power. Mbembe (2003) has forcefully expressed this concept in relation to African colonial sovereignty through the concept of 'necropower'; the government of death, not life. Operating in a state of exception and enmity towards African territory and peoples, colonial governments synthesised massacres and bureaucracy

Table 1.2   Colonial Indian governmentality

	Excess	Neglect
Sovereignty	Violence, ceremony	Democratised rights, centralised state
Discipline	Segregation, incarceration	Swarming, functional inversion
Economy	Exploitation	Withdrawal of state from economic processes
Society	Civilising ethos	Integration beyond the elite
Biopolitics	Knowledge creation, experimentation	Welfare ethos
Pastoral	Hyper-regulation of colonising society	Individualised care are for native population

long before the Nazi state existed. The difference of colonial governmentality was, thus, that violence became the language of right and exception
became the structure of sovereignty. Hansen and Stepputat (2005: 18–20)
agree that colonies can be thought of as near permanent states of exception,
defining themselves through the mercantile and military logic of exploitation, domination and civilisation. Yet, the production of bare life was just
one element of sovereignty; in addition there was the performance of public
authority and the marking of space through violence and rituals. In this process colonial territories became heavily impressed with the spatial insignia of
sovereign power, such as boundaries, hierarchies, zones and cultural imagin-
aries: 'Space was therefore the raw material of sovereignty and the violence it
carried with it. Sovereignty meant occupation, and occupation meant rele-
gating the colonized into a third zone between subjecthood and objecthood'
(Mbembe, 2003: 16). This violence was not only foundational, in that it
created the territory over which it was exercised, but also legitimating, as it
provided the model for colonial order, and imaginary, in that it embodied
the state (Mbembe, 2001: 25).

In appropriating Foucault's work on biopower, Mbembe (2003: 7) placed
racism at the centre of colonialism's distribution of death and life. Besides
violence, race is most commonly invoked as the essence of colonial difference (Chatterjee, 1993: 14). Yet, while race may provide the over-arching
dichotomy of the colonial episteme, Scott (1995: 195) has emphasised
the differences in how race was articulated in subject-constituting social
practices and how race cross-hatched with different forms of 'othering'.
As such, while colonial governmentality brought forth race as an object of
governance, this race was by no means constant (Hussain, 2003: 30).

While race was articulated with sovereign power in terms of violence,
sovereignty and race were also imbricated across the range of biopower.
Hansen and Stepputat (2005: 5) have insisted that colonial sovereign power
was more dependent on spectacles, ceremony and violence than that of
European centralised states. Thus, colonial sovereignty was characterised
not just by states of exception, but also by indirect rule from a distance,
asserted racial superiority and an oppressive governmental apparatus. In
terms of another technique of sovereignty, Hussain (2003: 32) and Howell
(2004b) have argued that the colonial sovereign state was 'full of law',
which infused every day practice. Yet, Hansen (2005: 176) has shown
that colonial law in India divided offences in line with assumptions about
caste and community, embedding the laws within wider governmental
programmes.

The translation of *discipline* was an ambivalent one. Outside of the liberal restraints of Europe, disciplinary institutions could experiment with
techniques of enclosure and segregation, as witnessed in the penal system
(Sen, 2000), the police (Sengoopta, 2003), the regulation of prostitution

(Ballhatchet, 1980; Levine, 2003) and urban reconstruction in the name of political security (Gupta, 1981; Oldenburg, 1984). Yet, at the same time, there was also an unwillingness to invest in the swarming of such expensive institutions (Gregory, 1998; Howell, 2004a). The result, thus, tended to be a tight and intense archipelago of institutions to protect the elite, while the rest of the population was left to the more distanced normalisation of colonial government.

It has been suggested that these actions of colonial *government* marked a stark discontinuity from Western norms. Prakash (1999: 125–7) argued against the idea of a tropicalisation of Western norms in favour of a fundamental dislocation, drawing on the case study of colonial India. Despite this, he uses the analytical categories of governmentality to dissect the difference of colonial Indian society with great acuity. Prakash cites the inability of the colonial states to forge a civil society, its need for despotic governance, its inability to mobilise capillary forms of power and its failure to respect the autonomy of interests outside of the state, in the liberal tradition. The resulting apparatus accrued detailed statistical understanding about India, although the tilting of government towards domination, rather than open power relations, meant that interventions in the disciplinary and governmental realm served as more obvious acts of colonial rule. As Rabinow (1989: 277) has argued, French colonies served as laboratories of modernity in which extra-metropolitan powers were experimented with. While the rhetoric of improvement was deployed, the colonial situation was in fact one of false *fraternité*, the denial of *egalité* and the absence of *liberté*.

This also helps to explain the relative absence of pastoral origins in the genealogies of colonial governmentality. There was very rarely an ethic of caring for *omnes et singulatim* ('all and one', Foucault, 1979a [2001]: 298–325) outside of the boundaries of the colonisers, and even they were more often subject to a disciplinary gaze rather than a confessional and personal interest. For native populations, the welfare ethos was only introduced in the British Empire in the twentieth century under nationalist and international pressure. This is indicative of a more widespread limitation in the activities of the state as played out through the domains of government.

In terms of the social, Kalpagam (2002) has shown that the creation of a colonial public sphere to encourage education and socialisation reverted to a disciplinary silencing when the colonial subjects asserted their own voice. The degree to which this public denoted the 'social' in the European sense has been debated. Prakash (2002) has insisted that the social was striated by religious community, while Chatterjee (2000, 2001, 2004) has argued that the social was divided into an elite, civil society and a political society for the rest of the population (see Chapter 4).

Colonial biopolitics was also heavily weighed upon by the demands of sovereign power for control and knowledge about the subject population.

Censuses and land surveys were implicated in the acquisition of knowledge about the 'disposition of things' and the nature of the multiplicities that had to be governed, although more intimate details were often collected than would have been allowed in liberal Europe (see Legg, 2006a). As such, colonial medicine functioned, and was seen as functioning, as a colonisation of Indian bodies (Arnold, 1993). In highlighting the ways in which Foucault's ideas on biopower had to be translated to analyse colonial Africa, Vaughan (1991) stressed that colonial biopower was more repressive than modern, focused on aggregate ethnic groups more than individuals and favoured generalisations about an already pathological colonial 'other' to detailed medical knowledge.

As a fourth element of translation, Vaughan also stressed that colonial capitalism was less modern and more obviously extractive than European forms, leading to underdevelopment. As part of a capitalist world system, and exploitative empires, colonial economies were often focused on creating cheap exports that could benefit manufacturers in the imperial heartland, while exposing colonial manufacturers to the competition of cheaply produced, industrial goods. This tended to lead to underdeveloped economies without the full spread of economic services and functions that could create robust and integrated economies (Goswami, 2004). However, as the colonial state created an Indian 'economy' through inducing new relations between resources, population and discipline, the emergent practice of economics also allowed nationalists to create an account of financial exploitation in colonial India (Kalpagam, 2000: 420).

As such, colonial governmentality, in India specifically, was subject to a series of excesses and neglects that can serve as a guide to the necessary translations of governmentality to the colonial context. These excess/neglects can be summarised as in Table 1.2.

Prakash (2002: 88) suggested that colonial governmentality's violation of Western norms means that no elegant triangle can be forged between colonial sovereignty, discipline and government. On the contrary, imperialism involves a translation of each of these forms of power to the colonial context, and then a site-based adjustment in relation to the most active 'nodes' in the triangle of power relations. Sovereign power was excessively violent and theatrical, but sovereignty was not devolved to regional governments or individual voting rights. Disciplinary institutions were excessively carceral yet failed to swarm through society or invert their influence onto the wider population. In general, the translation of the modes of power reveals a form of rule that put governmental apparatuses in place, but which had fundamental doubts about the ability of colonial populations to support the processes on which liberal government relied. The economy was thought too underdeveloped to be left to the forces of the free market and was thus heavily intruded upon by the state, although with the minimal investment to yield the highest profit. Society was thought too irrational and traditional to support representative institutions and too lacking in potential to justify wide-scale educative and/or social programmes. Biopolitically, the state sought knowledge about the details of the multiplicity of peoples within its territory, yet refused to finance welfarist interventions that would have improved the lives of its subject peoples.

That is, colonial governmentality was more an art of government than a science. It remained wedded to the apparatuses of regulation rather than security, to a model of police rather than one of liberalism. The Government of India remained too unsure of its security to rely upon the semi-autonomous processes of society, economy and population, and thus constantly sought to organise the disposition of things, yet with minimal investment or intrusion into the hallowed ground of 'tradition'.

The temporal differences of colonial governmentality

The colonial differences outlined above have mainly concerned power and space; the interflow of governmentalities between core and periphery. However, colonial governmentality varied greatly over time. Mbembe (2003: 17) stressed that late-modern colonial occupation was unique in its combination of the disciplinary, biopolitical and the necropolitical. Indeed, the evolution of colonies over time has tended to more closely imbricate the governmental tripartite. As such, Darwin (1999) has suggested that late-colonial states, and the negotiations they provoked, display at least six common characteristics. In terms of economic and biopolitical rationalities, the state became more *developmental*, by which the economy was modernised and the government of the population opened up to 'expert' influence. State institutions also became *denser* in a belated attempt to create, infiltrate and conduct the 'social'. But the state's sovereignty also underwent a series of shifts. The state became *bigger*, in that it sought to map and control all its territory, although often lacking the resources to do this effectively. Disciplinary and sovereign powers were forced into an alliance to create a *secure* state in the face of anti-colonial nationalism, while state sovereignty itself actually started to crumble. Sovereign boundaries became permeable as the state became more *open* to external influences. This foreshadowed the eventual emergence of the *self-destruct* state as it geared itself towards independence and the transfer of power.

The development of the Indian complex of sovereignty, discipline and government was, of course, incredibly intricate and could not be covered in any detail here. What this complex does show us very clearly is that colonial governmentality by no means reinscribed all earlier forms of sovereignty and power within its own framework (Dean, 2002c: 123). This was in part because the establishment of permanent administrational and territorial control with fixed borders and centralised administration came late in India's colonial history. As such, the colonial state always existed in a state of externality to an Indian society that was deemed unable of constituting or regulating itself (Prakash, 2002: 82). Even the laws that were introduced in an attempt to guide the subcontinent from oriental despotism to a utilitarian and civilising government failed to penetrate and reconstitute society beyond the institutions that were able to immediately enforce them (Hussain, 2003: 39).

In addition to the externality of the colonial state, previous forms of Indian state formation and government must be taken into account. Traditional Hindu society was not hierarchically organised under a strong state but was decentred and asymmetrically hierarchised in relation to locally stable socio-cultural systems (Kaviraj, 1994: 29). Hansen (2005) has charted the evolution of these forms of sovereignty as Muslim influence spread in the subcontinent from the ninth century. The Mughals invaded India in 1526 under Babur and, within 200 years, had established their authority over most of the subcontinent from their capital at Delhi (Blake, 1991; Hintze, 1997). Mughal forms of sovereignty, which encouraged greater professions of loyalty to the Emperor, imbricated previous forms of sovereignty. They were also forced to adapt to the encroaching sovereign powers of the East India Company (EIC) as it expanded westward from its base in Bengal in the eighteenth and nineteenth centuries, most notably through the Wars of Annexation (1793–818). It was only in 1803 that Delhi, still the Mughal capital, was taken by General Lake (Spear, 1973).

The EIC utilised forms of sovereignty such as local landholders, courts and religious authorities while slowly extending state-like institutions in a tentative technology of colonial government. Yet, as an imperial ideology began to develop within the British Empire, utilitarian policies began to be introduced to India in the 'Age of Reform' (1828–35). Rail, road, postage and telegraph infrastructures were strengthened, the army was reorganised, education policy was rethought and uncivilised India traditions, such as *sati* (widow immolation) were outlawed (Wolpert, 1977). This acceleration of the governmentalisation of the EIC state, elsewhere founded on excesses of colonial violence, did not have too drastic an effect on Delhi where, as a relative outpost, British residents were left to negotiate the relicts of the Mughal state in a period known as the 'golden calm' (Bayley, 1980). Yet, the movement of the cantonment outside of the city represented a growing distance between the colonising and the colonised societies (Stein, 1998). The British failed to penetrate Indian information systems and thus failed to anticipate the violent 'revolt' (referred to as the Mutiny in imperial historiography, the First War of Independence in nationalist historiography) of

1857 (Bayly, 1996). On 10 May Delhi was taken by the amassed forces and was only recaptured after a three-month siege (Dalrymple, 2006).

The bloody aftermath of the revolt served as the founding act of violence and the mythic origin of the *Raj*, established by the Government of India Act of 1858. After extensive executions and plunder, the entire population of Delhi was evacuated, only to be readmitted after the brutally cold winter season. Over the following years, a third of Delhi's urban landscape was destroyed in order to obliterate the communities that had supposedly conspired against the EIC, and to open the landform up to surveillance and the free movement of the military technology of violence (Gupta, 1981; Hosagrahar, 2005). The revolt offered the final proof that India could not be left unsupervised, with the freedoms of a liberal state. The belief in the 'Sameness' of the Indian population was shattered and largely replaced with the direct rule of an irrational 'Other' (Metcalf, 1994). While a cooperative native elite was crafted into a civil society that could supposedly bridge an increasingly aloof state and an increasingly agitated population, the rest of the population *was* targeted, but through the policies of political society that indirectly manipulated the population through apparatuses of security. Control was exerted over sanitary systems, taxes, public works, burial grounds, housing design and commercial areas in an attempt to improve circulation of air, water and capital. Yet, the population also experienced an increasingly brutal state without the benefits of liberal education or health care: 'The paramount aims of colonial biopolitics were to maintain stability and order, whereas the grooming of colonial quasi-citizens was highly selective and always circumscribed by both class and race' (Hansen, 2005: 177–8).

In terms of sovereign power, besides the continuation of military and police violence, the Victorian Raj was marked by racial distinctions, social hierarchies and excessive displays of the pomp and imperial ceremony, which saw Queen Victoria crowned as Empress in 1877 (Cohn, 1983). A celebratory Durbar was held at Delhi, attempting to tap into the city's ancient prestige, to be followed by a further Durbar, in 1902, celebrated the crowning of King-Emperor Edward VII. The last Durbar, attended by George V, was the ceremony of 1911 at which the capital transfer was announced and Delhi resumed its former status as capital of India. New Delhi would be administered centrally by the Government of India, with mundane duties eventually being passed from Delhi Town Planning Committee (DTPC) to the Imperial Delhi Committee (IDC), which later became the New Delhi Municipal Committee (NDMC). The rest of Delhi Province, an area of 547 square miles crafted out of the Punjab, was governed by a Chief Commissioner, Deputy Commissioner and the Delhi Administration, based in the Civil Lines to the north of Old Delhi (Chopra, 1976). These cooperated with the Delhi Municipal

Committee (DMC), which was presided over by the Deputy Commissioner and attempted to bridge local anti-colonial sentiment and the local administration. Under mounting nationalist criticism (see Chapter 3), the Government was forced to cede certain powers through the Government of India Acts (1919 and 1935) (Bose and Jayal, 1997), although no power was devolved to the Delhi Administration. The latter reform came just 4 years after the inauguration of New Delhi, which was eventually completed in 1931.

New Delhi: Showcase of Sovereignty

There are certain general principles governing town planning in all countries and climates, though they must vary with the motif of the city. First and foremost among these the Committee put the need of foresight ... Whatever eventualities the days to come may have in store, the new city must have at its hand the inherent power to command health, and a wealth of air spaces and room for expansion, which no lapse of time can deplete ... There must be beauty combined with comfort. There must be convenience – of arrangement as well as of communication. The main traffic routes must be parkways capable of extension both in width and length ... Where possible, there should be presentation of natural beauties – hill, wood and water – and of monuments of antiquity and of the architectural splendours of modern times. Space is needed for recreation for all classes. The result must be self contained yet possessing a latent elasticity for extension. The perfected whole should be obtainable with due regard to economy.

To all this must be added the special principles governing the town planning of a particular site. In the case of Delhi the Committee conceive the chief of these to be a realization of the dominant idea of the new Delhi and the adaptation of the scheme of the new city to physical conditions. Delhi is to be an Imperial capital and is to absorb the traditions of all the ancient capitals. It is to be the seat of the Government of India. It has to convey the idea of a peaceful domination and dignified rule over the traditions and life of India by the British Raj.[2]   (Delhi Town Planning Committee, 1913)

The transfer of the capital of British India from Calcutta, in the northeast of the subcontinent, to the more centrally located Delhi was not just motivated by location. It sought to remove the headquarters of the Government of India from the increasingly revolutionary province of Bengal, and to tap into the traditions and mysticism of Delhi, which had historically played host to 11 capitals, dating back to the construction of Indraprastha (c. 900 BCE). The transfer was undeniably a top-down decision, proposed by Viceroy Hardinge and a handful of close colleagues to the King in a confidential memorandum. The design of the capital was delegated to the DTPC, headed by

Captain George Swinton and including Edwin Lutyens, the future chief architect.

The DTPC had to negotiate a series of competing claims, and in so doing created a unique hybridisation of the imperial and the modern. As the quotation above denotes, the new city would embrace the emergent art of modern town planning. It would embody progress and foresight, anticipating and providing for the future, while combining functional comfort and economy with beauty of form. Yet these seemingly universal principles of urban government had to be tailored to the site, and to the political context. The new city was not just to be a colonial city, one of government, administration and bureaucracy. The working name for the new city was initially 'Imperial Delhi'. This denoted the centrality of the city to the twentieth-century aesthetic of British imperialism and its performance of showcase imperial sovereignty for an increasingly aggressive nationalist audience.

While this book will highlight how the capital project failed in many senses, it undeniably succeeded in its showcasing mission. The Viceroy's House and All-India War Memorial Arch by Edwin Lutyens and the Secretariats and Legislative Assembly by Herbert Baker have been near universally acclaimed. Amongst other things, this was for balancing the assimilation of local cultures within the civilising influence of Western architecture, and accommodating Indian traditions through the uptake of indigenous architectural forms.

As such, New Delhi can be read as a space of sovereignty in at least three respects. Firstly, the capital transfer marked an undemocratic decision by the ultimate authorities in the Empire to construct a new city. Secondly, the landscape aesthetic of the city represented the 'peaceful domination' of the Indian people. Finally, New Delhi has masterfully exerted its sovereignty over colonial urban historiography, establishing itself not only as a landform of utmost academic importance, but also as a self-contained city with sovereign boundaries and a clear distinction from the neighbouring city of 'Old Delhi'. Yet, the relationship between the new and old Delhis was not just one of separation, but also one of eclipse. This was a local and historical process in terms of prestige, financing, policing and improvement. Yet, the shadow cast by the new capital has also fallen upon colonial and nationalist historiographies, as well as architectural and urban works of research. As a space of colonial violence and display, or a site of nationalist resistance, mobilisation or factionalism, Old Delhi has been distanced and silenced. It is chained to a binary that depicts it as subordinate, Old and Other, against the powerful, New, colonial Self of the capital.

This silencing is a by-product of broader trends in historical and theoretical literature. The historiographical tradition of colonial urbanism has often supported the 'dual cities' hypothesis, encouraging the study of European settlements that bordered, and even intruded upon, native settlements

(Abu-Lughod, 1980, 1965). This focus has often eschewed the more wide-spread and insidious means by which colonial influence pervaded the native urban landscape (Yeoh, 1996; Çelik, 1997). This trend, in turn, feeds into, and has emerged from, the more theoretical tendency to study the policies of governments in abstract terms that deny the specificity of place, resistance and international context and fail to provide opportunities for political cri-tique. The following two sections will mobilise a Foucauldian approach to the spatial relations of Delhi to counter these two shortsights. The existing literature on Delhi will be reviewed to highlight the need for an approach that is cognisant of, yet not dominated by, the sovereign authority of New Delhi. Second, the local commentary on the inauguration ceremonies of 1931 will be reviewed to highlight the immanent potential for critique within the city itself.

A case for urban regicide? Beyond the capital

> We need to cut off the King's head: in political theory that has still to be done.
> (Foucault, 1980: 121)

In adopting a Foucauldian approach to power relations, this book will seek to explore power in its various guises, from the disciplining of individuals to the regulation of the population, removing attention solely from sover-eigns who detract resources from a position of externality to their population and territory. This politico-philosophical regicide will have a geographical corollary in Delhi, diverting attention away from the capital and into the urban capillaries and habitats of the two cities. Yet, the forceful presence of New Delhi's imperial landscape stands as a reminder that power relations cannot be conceived without sovereign power. Because of the lasting allure of this latter power, the literature on Delhi has mostly failed to divert atten-tion away from the showcase buildings at the heart of the city, or onto the older city beyond the capital.

King's (1976) study of New Delhi avoided the pitfalls of many later studies by focusing on the symbolism and layout of the city as a whole rather than the central complex. Within the geometrically aligned street roads, most of which were at 30°, 60° or 90° to the horizontal monumental axis of Kingsway, King analysed the hierarchical grid of social stratifica-tion along which government employees were distributed within the city (see Chapter 2, Figure 2.1). The bungalow compounds and their allocation formalised pre-existing social norms, as King (1976: 244) stressed: 'Distinc-tions hitherto informal or unarticulated were now clarified in the ordered physico-spatial divisions of Delhi.' King himself drew on the substantial work of Thakore (1962, see also Bopegamage, 1957; Mitra, 1970) who had studied the layout and history of the New Delhi plan, and devoted a large section of his work to analysing the spatial segregation of social class within the city. These patterns were also evident in the temporary capital to the north of Old Delhi, in which the government was housed from 1912 to 1926 while New Delhi was constructed. Such patterns necessarily disintegrated in Simla, the summer capital in the foothills of the Himalayas to which the whole government annually migrated. Following King, many commentators have summarised the residential divisions as part of what Davies (1985: 274) called the 'petrification' of Edwardian India in the urban landscape (also see Sealey, 1982).

Yet, following Irving's (1981) exemplary study of the capital, successive authors have focused on the architectural heritage and design of New Delhi. Vale (1992: 56) has suggested that the continuity of themes and designs in capitals including Washington DC, New Delhi, Canberra, Ankara, Chandigarh and Brasilia forms a 'hermetic dialogue'. Jain (1990) has even attempted to link the alignments and functions of the city to the ancient Indian Vedic texts, although they were actually dictated by Viceroy Hardinge, aligning Kingsway with Indraprastha Fort and Parliament Street the Jama Masjid in Old Delhi (Nilsson, 1973: 54). Davies (1985: 215) suggested that the relevant context for New Delhi was that of the high British Imperialism of the late Victorian period, which called for a common architectural language to unify the Empire and represent Britain's strength.

Other studies have focused on the nature of this representation in New Delhi. Metcalf (1989) has stressed the repeated calls to include traditional Indian styles and craftsmanship in the capital, while Volwahsen (2002) has comprehensively traced the architectural genealogy of the city. Ridley (2002) has detailed the personal politics between Baker, Lutyens and Hardinge who favoured, respectively, the interweaving of Indian features within the narrative fabric of imperial architecture, the elemental and classical architectural style, and an Indo-Saracenic compromise. While Stamp (1981) and Ridley (2002) have insisted that, in the 'Battle of Styles', Lutyens succeeded in synthesising the east and west, Tillotson (1989: 122) maintains that eastern features serve only as 'punctuation marks' on an essentially Western building (see Hopkins and Stamp, 2002, for a contextualisation of Lutyens's international work).

Another aspect of the New Delhi literature has stressed its ceremonial and ritualistic spatiality (Hosagrahar, 1992). Viewing the landscape in its performative dimension, Christensen (1995: 43) argued that the opening ceremony of New Delhi in 1931 marked an evolving imperialism in which Dominion status was under debate. This was an Empire at the beginning of the short twentieth century, not the end of the long nineteenth. As such, Hall (1988: 177–88) categorised New Delhi as a 'City of Monuments' and as part of the 'City Beautiful' movement, alongside Chicago, Berlin and Moscow. Likewise, Dalrymple (1993: 82) links the ceremony, inhuman scale and racially superior ideology of New Delhi to Fascist Italy and Nazi Germany. Less extremely, Stamp (1981: 40) stresses the modernism of the capital, relating it to the industrial and the commercial.

All these texts are indebted to Irving's (1981) pioneering work, yet many failed to stress the ambiguities and tensions within the capital as he did. Lord Curzon, the ex-Viceroy, vociferously opposed the move from Calcutta, the trading communities in Bengal bemoaned their loss of influence, there was continued debate regarding whether the city should be located north or south of Old Delhi, while Viceroy Hardinge was nearly killed in an assassination attempt during his State entry into Delhi in 1912. The time span and cost of the project spiralled out of control while the Great War drained resources from the capital project. Most famously, Lutyens and Baker consistently disagreed over fundamental elements of the city's design leading to a series of furious rows.

However, Irving's study is conducted in a near vacuum regarding Indian politics, with the growing nationalist movement only being mentioned in the conclusion (Rabinow, 1983). Similarly, Old Delhi is only mentioned as a distant referent. A few authors have rectified this pattern. Morris stated that

> New Delhi was an anomaly – too late for arrogance, too soon for regrets, too uncertain to gets its gradients right ... The city lacked both the insolence of conquest and the generosity of concession, and by its deliberate separateness it perpetuated invidious old comparisons.   (Morris, 1983: 221–2)

Such comparisons included those of death rates, which in the old city were four times that of the new. Architectural comparisons between the two cities have also been made. Sorkin (1998: 67–8) commented that New Delhi was designed as the inverse of the winding streets of Old Delhi, while Evenson (1989: 148) suggested that the possibility of the two cities visually harmonising was never seriously considered (although Lanchester's original plans for the city proposed just that). Evenson summarised the attempts to improve the old city (see Chapter 4), while Chatterjee and Kenny (1999) have emphasised how these works related to the new capital. Despite these few examples, the overwhelming impression given in the literature is that there was only an insignificant measure of interaction between the New and Old Delhis. Yet, Gupta (1981) has shown that before, and during, the construction of the new capital the old city was a space of intensive governmental violence and regulation, as Hosagrahar (2005) has charted through the architecture of the city.

In fairness to those who have written on New Delhi, they are situated within a much wider historiographical trend that has fortified the dual city divide of the colonial/native urban form (see the concluding chapter for a discussion of recent colonial urban studies that buck this trend). Çelik (1999: 374) has noted that many studies of 'non-Western cities' attach them to binaries in which they are denied autonomy. As she stated, 'Behind the clear message conveyed by the image of dual cities at first sight, however, hide more complicated implications' (Çelik, 1997: 5). King (1992: 341) has, likewise, stressed that (post)colonial urban studies still render indigenous cultures as 'traditional', which epistemologically prolongs the original colonisation. This should be met, King suggests, with an emphasis on the colonised and the subaltern, seeing the colonial city as what Yeoh (1996) terms a contested terrain of daily routines and conflicts.

There is also a more fundamental tension regarding power relations that is being brought into the dialogue here. Butler (1997: 2) has articulated this tension as the paradox of subjection. Firstly, power often appears as dominance by an external force; that which subordinates and relegates lower orders. But, following Foucault, Butler also suggests that power constitutes and provides the conditions for existence of those who are subjected. While power *does* press down on individuals through certain techniques or strategies, it also infiltrates, creates and alters the constitution of the subjects of power. This is very much a spatial process, and the geographies of this power (Allen, 2003) remain lost for many of the 'native' or 'local' cities that were drawn into colonial power relations. Butler's theories on performativity and subjectivisation suggest one way of considering these geographies, yet her failure to maintain an emphasis on local, material power relations means that she misses out on many of the main points of instability that animate social practices (Mills, 2003). Similarly, Said's (1978) combination of Gramsci's coercive elements of hegemony with the Foucauldian discursive production of complicity presented a theoretical solution, but one that Said (1986) himself rejected on the grounds of Foucault's anti-humanism, his apparent apoliticism and his failure to theorise resistance. The continued engagement of post-colonial scholars, and others, with Foucault's later work suggests that Said's pessimism was not justified, while the recently translated lecture courses of Foucault are shedding light on the degree to which sovereign power intersected with biopower in modern governmentalities. As such, the task is not one of removing the capital, of urban regicide, but of situating sovereignty within the dense network of urban power relations upon which it depended.

The tombstone of the Raj?

Although the emphasis of this book will be upon the government's ordering of colonial space, this took place within the context of an emergent and highly effective anti-colonial nationalism (as explored in Chapter 3). The Rowlatt disturbances of 1919 marked Mahatma Gandhi's first mass movement, which was followed by the campaigns of Non-cooperation (1920–2),

Civil Disobedience (1930–2) and Quit India (1942). These movements created and capitalised upon a groundswell of discontent and rejection of the colonial government, which not only saturated the old city but also penetrated the New Delhi. Yet the capital also revealed itself as an aporetic crisis object in the field of representation (Shields, 1996). Irving (1981) has shown how the design, construction and cost of the city were all criticised. Yet, there was also a mixed local reception at the inauguration of the city that pre-figured its problematic operation as a fully functioning city.

Many of the statements produced in appreciation of the city during the inauguration unintentionally echoed the wishes of the DTPC, as is evident from the quote that opened this section. Viceroy Irwin's speech on 10 February 1931 echoed the need for *foresight* and professed a desire to protect the city from the evils that accompany city growth. The *economy* of the city was commented upon, although only to marvel at just how much the city had cost; the £10 million price tag being way in excess of the original budget.[3] The *imperial* nature of the project also occupied a number a commentators. A special edition of the *Indian State Railway Magazine* commented that the city had managed to combine east and west, amalgamating influences from antiquity to the Mughals (Shoosmith, 1931). The *Daily Herald* referred to it as 'A dream city of the East, in which is mingled the best of the west …'[4] Yet the city was more widely described in terms of the West surpassing the despotic and collapsed cultures of the East. In an article for *The Bengal, Bihar and Central India Annual*, Mrs Shoosmith, the wife of an architect involved in the project, classified all past capitals in Delhi as military despotisms, while the new city hosted representatives of India's constitutional government.[5] The approval of New Delhi's representation of the new sovereignty continued:

> On a massive foundation of red, fortress like, rooted in Indian tradition, stand the white columns and walls of palaces of an age of greater enlightenment … Now darkness shrouds the older Delhis and engulfs the historic plain; while, strangely and dramatically illuminated from below, the great dome of the palace, like some gigantic presiding genius, broods over the new City. (*The Bengal, Bihar and Central India Annual, 1931*)

In terms of the DTPC's 'special principles' for Delhi, the British press reserved attention for the capital's function as the seat of government. The *Daily Telegraph* suggested that the city would appeal to the legends of the Indian past, reinforcing the permanence of British sovereign rule,[6] as coordinated in what the Canadian representative at the inauguration ceremony referred to as a 'temple of government'.[7] This function was intimately connected to what the DTPC referred to as the 'peaceful domination' of the Indian people. *The Observor* commented that the prophesy that any Empire which located itself in Delhi would be lifted because '[t]his is the end of the old Empire, and its transformation into the British Commonwealth of Nations at the Crown's own initiative'.[8] *The Guardian* also looked optimistically towards the future, reading New Delhi not as a vainglorious gesture of domination but as the home from which India would plan her future.

There was, however, a counter-discourse that stressed the more negative aspects of this showcase of imperial sovereignty. Architectural criticism of the city had continued since the opening debates regarding the designs. The Government was forced to respond in the Legislative Assembly in 1927 to reports that New Delhi was ugly, unoriginal, of unimaginative pomposity, monotonous mediocrity and was more in the nature of a prison than a habitation.[9] *The Times Book of India* (1930: 161) pointed out that the accommodation provided in New Delhi was totally inadequate for the government's clerks and that they had been forced to add to congestion in Old Delhi.

Such feelings were further provoked during the inauguration, which took place just a month after the suspension of the Civil Disobedience campaign that had been led by Gandhi since March 1930. As such, the disciplinary actions of the state were highlighted in the *Hindustan Times*' editorial of 13 February. Viceroy Irwin was mocked for suggesting that the bonds of the empire were 'freedom' when thousands of people were still in jail for attempting to exercise that freedom. The state's government in the economic domain was also criticised because the 'repressive regime' had worsened the effects of the Depression, making the lavish inauguration celebrations an insult to the Indian people.

The ethos of the celebrations was also criticised: 'The whole outlook of the proceedings was imperialistic and gave one the impression of having been designed to demonstrate the supremacy of the White man. An underlying strain of imperialistic sentiment was present throughout His Excellency the Viceroy's speech.' The celebration was said to ignore Indian sentiment, which was still despondent at the repression of Civil Disobedience and was mourning the death of nationalist leader Motilal Nehru. British papers were also critical. *The Times* decried the lack of popular support and the plastering of the city with armed police who stamped out any demonstration.[10] Picking up on this depressed atmosphere, the *Yorkshire Post* commented on 11 February 1931:

> A solvent of ancient griefs, a cement of new loyalties, an earnest of cooperation – these, it may be, were the hopes chiefly placed upon the conception of Imperial Delhi when the project was formally proclaimed ... Yet the situation has changed ... So New Delhi is inaugurated in an atmosphere of political uncertainty rather than of political confidence. There are in India those who see in it a memorial, indeed, to British enterprise and orderly development, but also a sepulchre of British influence and authority in India. It appears to them the tombstone of the British Raj.

The contrast could not have been greater than when Gandhi arrived in the city a week later. As opposed to the attendance by invite at the inauguration, and the muted celebration of an antiquated imperial aesthetic, Gandhi was rapturously received by an estimated 60,000 people in Old Delhi. Many of these followed him to the Viceroy's House in New Delhi where he negotiated terms with Viceroy Irwin.[11] This performance of the avatar of anti-colonialism within the heart of the imperial capital represented a more thoroughgoing resistance to, and failure of, the imperial project that had been gradually developing within the capital city, as demonstrated in the following chapter.

Chapter Two

Residential and Racial Segregation: A Spatial Archaeology

The emphasis in post-colonial studies on the Other has been reflected in other disciplines that focus on identity politics. Such studies have highlighted oppressed communities, marginal spaces, ethnic minorities and stigmatised sexualities. Yet, there is also an acknowledged need to focus on the construction and mutual constitution of seemingly hegemonic identities of heterosexuality, the middle classes, masculinity, or whiteness. Although an emphasis on agency and resistance is necessary, Duncan and Duncan (2004: 27) were right to stress that an over-emphasis on these achievements '… can sometimes deflect attention away from a critical, grounded analysis of the workings of successful hegemony, structured inequalities, unintended externalities, unknown conditions, and complex complicity across far-reaching networks'.

In contrast to the aesthetic hegemony of the North American suburbs that the Duncans have studied, on close inspection, the landscape of New Delhi demonstrates something more akin to Guha's (1997) domination without hegemony. While the city's beauty was appreciated, the ethos underwriting the landscape was both obvious and overtly criticised, whereas the urban form failed to function efficiently. This does not imply that guerrilla warfare or the weapons of the weak brought down the city. Rather, the attempt to impose imperial order on the landscape through the housing *techne* was under-funded, met a recalcitrant and reflexive public, and did not keep up with the changing national context.

The first section of this chapter will explore the imperial *vision* of order for New Delhi and the *identity* assumptions that informed the concepts behind it, inline with the analytical approach summarised in Table 1.1. The second section will show how this ideal landscape was *problematised* by over-crowding and under-building. This situation created physical and discursive positions from which both individuals and institutions could criticise the government. These critiques highlighted the *ethos* that so obviously informed the landscape ideals of the Government of India, which responded with protective measures against, rather than alleviative measures for, those who could not find accommodation in the city. These studies will be explored using Foucault's archaeological methodology, which will also expose the colonial cultural geography that informs the other landscapes explored throughout this book.

Archaeology

Foucault established his reputation as an analyst of Western society's Others. His books concerned interpretations of the mad (Foucault, 1967), the ill (Foucault, 1973) and the criminal (Foucault, 1977). Yet, he also had an abiding interest in the majority, whether this was his governmental studies of regulation or his archaeological studies of ordering. His work in *The Order of Things* (Foucault, 1970) and *The Archaeology of Knowledge* (Foucault, 1972) studied not the Other but the Self, although he remained obsessed with the idea of difference (McNay, 1994: 48). This combination of sameness and difference comprises the ideal toolkit to analyse the united yet fractured landscape of New Delhi, and colonial society more broadly. Indeed, Stoler (1995: 143) has questioned why post-colonial studies have dwelled so little on *The Archaeology*, and so much on colonial governmentality. As will be shown, the two approaches are compatible and the tools from the archaeological works help us to understand the episteme not only informing the regime of landscaping in New Delhi, but also the discursive/spatial formations that infiltrated the older city to the north.

In *The Archaeology of Knowledge*, Foucault (1972) sought to provide methodological coherence to his earlier works. He emphasised his interest in discontinuity and the emergence of new rationalities that were, simultaneously, transformations and displacements of earlier concepts. He had most famously explored these shifts in terms of the *epistemes* of ordering words and things in Europe since the Renaissance (Foucault, 1970). Despite the grandness of this project, Foucault stressed that his aim was a general, not a total, history (Foucault, 1972: 10). This generality concerned the series, limits and shifts of particular epistemic forms, not organised *around* a common *history*, but *through* the *spaces* of dispersion (Philo, 1992: 148).

However, these spaces were those of discourses, those linkages of statements, knowledge and practice that determined what could be thought and said, what was 'true', in a particular time and place. The emphasis in archaeology is not on what is known or why knowledge is possible, but on how knowledge is ordered (Major-Poetzl, 1983: 21). The *Archaeology* marked a turn away from social practices to linguistic effects in Foucault's work. This turn would only be reversed in his later genealogical works (Dreyfus and Rabinow, 1982: viii). Yet, this focus on discourse produced a sophisticated methodological framework that can be applied to social contexts. Foucault (1972: 31–7) sought to link statements together and describe their connections through 'discursive formations'. Rather than referring to the same object, style, concept or theme, discourses are marked by their laws of division, co-existence of heterogeneity, emergence of new objects and the dispersion of points of choice. Their unity is in their system of dispersion, and this system can be analysed through its rules of formation. These rules are not rigid delimitations but denote recurrent patterns in discursive practice that can be used to explore the norms of social ordering.

The details of the four rules of formation that Foucault identified will be examined more closely when they are deployed as structuring devices throughout this chapter, but the basic framework is as follows. The rules mark the conditions of existence, maintenance and disappearance of a discourse. First, they refer to the *objects* of a discourse and the way in which they emerge in new registers. This emergence can be traced through 'surfaces of emergence', 'authorities of delimitation' and 'grids of specification'. Second, there are *concepts* that inform the wider discourse itself and describe its organisation. *Enunciative modalities* mark the subject positions of discourses and the status of these positions. Finally, *strategies* organise the concepts of a discourse into wider coherence and theoretical viewpoints. The emphasis on the local and lived nature of colonial discourse in this chapter means that the strategic play of discourses will not be examined at length. As Gutting (1989: 232) has stressed, not all of these rules have to be in evidence; it is the system of dispersion to which these rules direct us that is the substance of a discourse, and this system was incredibly strong in the New Delhi landscape. Here, the object of discourse was 'housing', not just as a material form, but rather, as a physical space of subjectification in which authorities of delimitation sought to impose utopian social orders on parcels of physical space and the subjects that occupied them. Housing is, thus, posed as a relational space in a grid of specification, not as an enclosed space or introverted home.

Foucault (1972: 208) predicted that his archaeological tools might later be dealt with as problems in a different way, and utilised alongside different methods. Both predictions have materialised, first, through a critique of *The Archaeology* and, second, through an application of its terms to the colonial context. In terms of the former, 'discourse' is criticised as being supposedly autonomous, yet simultaneously possessing causal efficacy, and as such as not being sufficiently theorised in relation to power: 'Moreover, it seems clear that the regularities [Foucault] describes are not simply accidental ordering which can be read off the surface of discourse, but that they must be evidence of some underlying systematic regulation' (Dreyfus and Rabinow, 1982: 84). It is the autonomy of Foucault's discourses that has unnerved so many reviewers, leading him to underplay the role of the material, the individual or the social. These questions will be addressed empirically in the second section. While acknowledging the internal heterogeneity and contradictions in discourses, Major-Poetzl (1983: 164–6) showed that Foucault's conception in *The Archaeology* eliminated non-paradigmatic formulations, failed to explain change and failed to allow individuals to craft their own subject positions.

It is because of these valid criticisms that the archaeological methodology must be put into practice alongside the genealogical works of Foucault's later life. These writings reversed the linguistic turn and emphasised social practice, placing discourses in relation to power. Indeed, in looking back on his archaeological works in 1977, Foucault asked what else he could ever have been writing about *but* power (Foucault, 1980: 115). In *The History of Sexuality*, Foucault (1979b: 92) spoke of discourses of sex explicitly in terms of power, and later in terms of self-formation (Foucault, 1986b), but he also returned to the archaeological complexity of discourse in reference to its 'tactical polyvalence' and contradictions (Foucault, 1979b: 100, 102).

While this turn towards power relations does mark a shift in Foucault's writings, Major-Poetzl (1983: 42) showed that in 1968 Foucault had defended archaeology as a progressive political perspective. He claimed that it established the limits and location of a discourse, thus enabling a realistic politics that did not pursue endless origins or hidden meanings. Yet, it was only with the genealogical emphasis on embodiment, materiality and self-government that such a politics became conceivable.

Some authors have shown how the archaeological and genealogical approaches interweave. Dean (1994: 34) has stressed that whereas an archaeology can be used to analyse local discursivities and highlight problematisations, genealogies focus on relations to the self. Laurier and Philo (2004: 424) refute Deleuze's (1988) suggestion that the archaeological work was distinct from later genealogical investigations. In a more empirical study, Hannah (2000) has used the archaeological framework to show how governmental rationalities emerged in nineteenth-century America. This chapter will argue that: the formation of new discursive objects provided new means of visualisation; that identity assumptions informed discursive grids of specification and enunciative modalities; that the conceptual landscape constituted a techne that bridged the material and the performative; and that the colonial ethos was present in the hierarchies that penetrated the whole city. This was within an episteme that has been the subject of colonial discourse analysis, although this vision of 'discourse' has been very different from the 'discourse' of *The Archaeology*.

Colonial Spaces of Dispersion

Said's depiction of colonial discourses creating a world riven by a Manichean divide has great force as a broad concept. However, various authors have complicated this divide at the level of interaction. From Bhabha's (1994) mimicry and ambivalence to Spivak's (1988 [2000]) silenced subjects, the negotiation of colonial discourse has been questioned. However, Young (2001: 399) has explored Said's very conception of discourse itself. Young suggests that Said conflated discourse and text, removing the importance of material context from the consideration of the former. Based on *The Archaeology* (Foucault, 1972) and later works in the *History of Sexuality* (Foucault, 1979b), Young insisted that discourse arose from practice, at the interface of language and the material world. While Said mistook discourse for representation, he also downplayed the heterogeneity and contradictions of discourse on which Foucault placed such great emphasis. These tensions could emerge within or between 'orientalisms' (Lowe, 1991), or as discursive displacement prompted by material change (Behdad, 1994). As such, this framework can be used to analyse the colonisers themselves (as urged by Rabinow, 1984: 201). This highlights the role of the imperial encounter in forging gender, sexuality, class and race identities, not in the core or periphery, but in the imperial circuits between imperial and colonial sites (Stoler, 1995).

Young (2001: 407) reserved further criticism for post-colonial theory's treatment of resistance and the subaltern. The idea that the subjective voice of the colonised faced the objective discourse of the coloniser eschewed Foucault's writing on enunciative modalities and his later assertions regarding power and resistance. Rather, colonial discourse led to a profusion of subaltern discourses (also see Phillips, 2002), although this did not constitute every possible position because of the birth of counter-discourses, such as nationalism, that operated beyond and around colonial discourse. For Young, a Foucauldian colonial discourse analysis would use discourse as a way of studying colonial practice within specific administrational regimes:

> Colonialism as a practice operated at the interface of knowledge and material culture, its operations were highly dispersed, contradictory and heterogenous in historical and geographical terms. Its discursive formations are likely to have been similarly heterogenous and subject to successive transformations in response to specific events.   (Young, 2001: 408)

In detailing how to study these discourses, Young drew attention to their *sites*. These could be European, including Parliaments, chartered companies or educative institutions, or colonial, such as administrative bodies, newspapers or memorials. These are testimony not just to the practices of colonialism, but also to its geographies.

Even in his work that did not refer to material places, Foucault's often couched his phrases in terms of space. For Major-Poetzl (1983: 23), Foucault's archaeological conception was inherently spatial, just as his other earlier works had been interpreted in terms of cartography, topography and geometry. However, we must go beyond Foucault as an abstract 'geometer of power' to examine the historical geographies his work brings to light (Philo, 1992). Since Foucault's earlier archaeological works and later genealogical works explicitly referred to social practice, their geographies were directly traceable (Driver, 1993; Ogborn, 1993a; Philo, 2000; Howell, 2004a). *The Archaeology*, however, requires the application of a methodology primarily intended for linguistic analysis to space. Duncan and Duncan (1988) showed how effectively this could be done through focusing on the work of Roland Barthes, although also referring to post-structural texts more generally. The assumed autonomy of discourse was criticised in favour of a more sociological analysis, as was the plurality of discourse in favour of empirical contextualisation. Thus qualified, a post-structural form of landscape analysis can unsettle the taken-for-granted assumptions about the 'natural' ordering of the environment and privilege individual readership of space. For instance, Duncan (1990) analysed the Ceylonese city of Kandy as a text, reading the religious and secular landscapes for the presence of power relations as allegory, synecdoche and metonymy.

As a strictly regimented space that conformed to more directly identifiable and complex rules of formation, New Delhi is more amenable to an archaeological landscape analysis. This included, but is not restricted to, what Mitchell (1988) referred to as 'enframing'. The term is from Heidegger, but the project itself was later enframed in Foucauldian terminology (Mitchell, 1991). Mitchell charted the spread of panopticism through colonial Egyptian agrarian land reform, educative institutions, town improvement and model housing. Such enframing of space sought to remake people as more visible and productive subjects. Space itself was also transformed, into something that is abstract; akin to frames or containers (Mitchell, 1988: 45). But these frames made people enumerable and thus enforced a social hierarchy. However, as Çelik has stressed in her comparable work on colonial Algiers: 'Because a form – even a seemingly crystalline one – can be viewed from a myriad of perspectives, focusing on physical aspects alone does not allow for a meaningful analysis of the city' (1997: 4). Çelik went on to focus on the urban process, although with the emphasis remaining on architecture. It is through this combination of enframing and perspective that the rules of discursive formation will be used to analyse the imperial, yet contested, landscape of New Delhi.

The Spatial Administration of Precedence

Visualisation: Objects of imperial discourse

New Delhi marked a break in British colonial housing policy. Whereas troops and bureaucrats had been housed in civil lines and cantonments before, this was the first housing project on such a scale, and to such a degree of rigid social segmentation. Foucault (1972: 42) stressed that the existence of new objects of discourse would be governed by several rules, one of which referred to the 'authorities of delimitation'. These were people or bodies who could delimit or identify objects of discourse and were defined separately to the enunciative modalities that emerged from within discursive formations. The original authority following the capital transfer was the Delhi Town Planning Committee (DTPC), which passed on its powers to the Imperial Delhi Committee (IDC) on 25 March 1913, which then became the New Delhi Municipal Committee (NDMC) on 3 March 1927.[1] The DTPC produced its first report in June 1912, which discussed the site for the city and the major engineering and health challenges.[2] The second report, from March 1913 dwelt on the design of the city and its layout.[3] These designs addressed not only the principles of the colonial modern, as discussed in Chapter 1, but also the layout of the city. It was made clear that Indian clerks would live near Paharganj, bordering Old Delhi, European clerks would be to the south of these, whereas the Commander-in-Chief would live near the Viceroy, around whom the officials of the Government of India and the Members of Council would cluster (Figure 2.1).

The imprint of the DTPC's ideals were left not only in the geometric layout of the city, but also in the ideas of its aesthetic, politics and zoning. The latter would prove to be a key 'surface of emergence' in which individual difference emerged, was designated and was then analysed. Zoning separated individual inhabitants from their social context and abstracted them to the level of social functionality; what they did for the government and thus what they deserved in material benefits. As Hannah (2000: 43) commented, these surfaces need not be linguistic. They are also the institutional sites in which objects become *visible*. This is where they are constituted as discursive objects themselves in time and space. In terms of the latter, places are designated with a homogenous function, to which individuals can later be allotted.

King's (1976) foundational study established New Delhi as the paradigmatic example of colonial urban residential zoning, dependent as it was on the 'conceptual' hierarchies of the Warrant of Precedence (see the following section on the conceptual landscape). This core principle of the city did not just emerge from the architects in the DTPC. They themselves were subject to higher authorities. For instance, during the negotiations in 1912 regarding which chief architect would be selected, Geoffrey de Montmorency, the

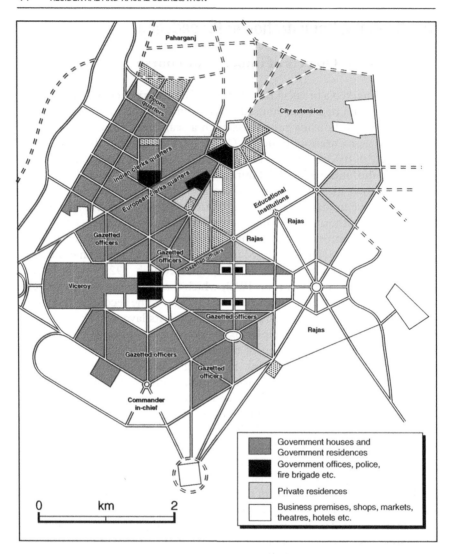

Figure 2.1   Sanctioned layout of New Delhi and the use of land as in 1914

Personal Assistant to the Chief Commissioner, annotated an alternative city layout. He included a note '... showing how all areas, which we require, would fit in detail into this layout, and the principles on which it has been attempted to group them. The same ideas would apply, *mutatis mutandis* [with due alteration of details], to any lay-out'.[4] De Montmorency went on to become the Secretary of the IDC and, in a project estimate of 29 December 1913, claimed that the aim of the allocation of areas for various classes was '... to assess the convenience for each class, while ensuring broadly a fair distribution over all classes of the general amenities of the site.'[5] It was suggested at this time that most clerks would be employed in the offices that would be located near 'Railway Place' next to Old Delhi, which became Connaught Place. They were thus accommodated in the north of New Delhi, adjoining the old city. The suggestion that only a minority of the clerks would be required to work in the Secretariat at the heart of New Delhi was a bizarre one, and would be greatly criticised when the city was occupied.

As such, the residential landscape of Delhi formed a surface of emergence in which relational objects of discourse were formed: functional housing. However, the administration of the landscape also provided the surface for the emergence of a connected object of discourse: class (see Chapter 3 for details of how these class geographies were policed). This will be addressed, first, in the gradational sense, not in relation to production or class consciousness. However, this fails to appreciate the complexity of the class concept, which will later be explored in relation to the 'grids of specification' by which the orders here described were actualised.

In *The Order of Things*, Foucault (1970: 53, see also Steinberg, 2006) stressed that beyond the establishment of functional places (distribution), ordering required a norm against which difference could be calculated (measurement). The norm need not be abstract, but can simply be the first term in an ordinal rank, which then becomes a hierarchy. The normalising factor in New Delhi was 'emolument' (salary). Each class of house had a wage bracket attached to it that dictated the class of the resident. However, the actual process of allocation conformed to a complex set of rules, which mediated the many competing claims for the best houses. The city's 'grid of specification', as discussed below, belied the notion of a single hierarchy for the city. This was vividly expressed by the existence of two similar, but separate, sets of housing allocation rules. One was for officers, who were almost entirely white in the 1920s when the rules were issued, and the other for clerks, the majority of whom were Indian.

The Secretariat staff finally moved from the Temporary Capital to New Delhi in 1926. In 1927, A.M. Rouse, the Chief Engineer to the Public Works Department (PWD), issued the 'Rules governing the allotment of quarters in New Delhi', divided into those referring to officers' houses and clerks' quarters.[6] The former rules also excluded those who were beyond the mundane hierarchies of the bureaucracy: Government House, the Commander-in-Chief's residence and the houses of Honourable Members or Presidents of the Legislative Assembly were all excluded. For those within the remit of the rules, the authorities of delimitation were clearly stated. The Estate Officer would allot and surrender houses, three Executive Engineers would repair buildings, supply filtered water and manage the electrics, while a Superintendent of Horticultural Operations would manage and water the compounds.

The classification of houses was explicitly linked to the gradations of income: from class A (above Rs 3,000 per month) to class D (below Rs 999 per month). Waiting lists were held for each class and to get onto the list an officer had to supply his name, post, date of appointment, emoluments and the details of any house desired. As such, the process of allotment was not entirely abstract. Previous occupation of a house established a 'lien' that would prioritise a claim over a newcomer to the city or someone recently promoted to the same class bracket. This went some way towards establishing a sense of place and continuity for a city in which the occupants were migratory, spending part of the year in Simla.

Lien was the strongest factor in application priority, but second came those men who could not get a house in their own 'class' and had to take a smaller house. After this the waiting list was allotted in terms of 'order of seniority', as established by the Warrant of Precedence, which ranked all Government employees from 1, the Viceroy, down to 66. The rent was confirmed as one-twelfth of the pre-stipulated annual rent, but not more than 10 per cent of an employee's salary. This money was subtracted from an officer's pay such that, to all effects and purposes, the house came with the job. Officers were given a copy of the rules, which dictated that the grounds could not be altered and had to be maintained, on pains of being evicted from residence and banned from future consideration. The clerks' rules were almost exactly the same, but had different class brackets. These ranged from A (above Rs 450 per month) to D (below Rs 224 per month).

The existence of two almost identical sets of rules is testimony to the degree to which the modern functionalism of urban space was combined with a colonial rationality that enforced the *ethos* of racial difference, here masquerading as class distinction. The seemingly democratic hierarchy was fractured by race at two levels. First, membership of the higher echelons of the Indian Civil Service (ICS) had been restricted to Europeans until the early twentieth century and it was still difficult for Indians to rise through the bureaucratic ranks. Second, the clerks' quarters were explicitly divided into European and Indian quarters. The division of the rules stressed that the city was not one seamless hierarchy but a divided city of governors and menials. Nor was the surface upon which class placement was allotted a smooth one. There were distortions and gradients, and the authority of delimitation could be contested by those within the system. For instance, the system dictated that two people on the same emoluments would have priority determined by the 'precedence'. Yet, in November 1929 it was suggested that the length of time the emolument had been drawn should be prioritised over precedence. The Army Department protested vociferously against this because the military were placed in positions of seniority in the Warrant, as the lowlier ranked representative of the Industry and Labour Department was quick to point out.[7]

This hinted at a deeper-seated tension regarding the basis of a gradational understanding of class. The original conception of class, embodied in de Montmorency's note, was that areas would be allotted to classes as defined by their role (precedence). Yet, the allocation of a type of housing to a pay bracket instilled a different form of landscaping by role (emolument). The Joint Secretary of the Home Department commented, on 27 November 1929, regarding the proposed changes:

> The present rules combine two principles: (1) emoluments, and (2) precedence. It seems a thoroughly illogical procedure that a man who has got into a class by reason of his pay, should then, when he is in the class, be ranked in it according to the warrant of precedence. This seems to me an unjustified and, indeed entirely wrong use of the warrant of precedence.[8]

The objection here was to the subordination of 'class' as defined from above, to 'class' as defined from below. The ranks of the warrant, of course, correlated to the ranks of pay. Yet, the ascriptive nature of the Warrant safeguarded certain positions from the more meritocratic, achievements-based class system of pay.

A further protest that had been raised in 1929 was that the system was too complicated to be administered centrally. As such, it was argued that block allotments should be made to individual departments of government, who would then allot the housing to their staff. Despite the Home Finance Department's complaint that constant disturbances in the system of classification were 'disturbing and bad', this suggestion seems to have been carried through. In March 1931 a question was raised in the Legislative Assembly regarding allocation of clerks' quarters in New Delhi.[9] On being asked about the waiting lists, the Government responded that allotment was made *pro rata* (proportionally) to each department, who allotted housing according to their own lists. An internal Home Department enquiry in response to a Legislative Assembly question in August 1934 indicated a system that had strayed substantially from the official rules.[10] Some departments allotted with regard to seniority, other to lien holders, others to married couples over single men and others with regard to proximity to their office. The Public Works Branch Department, who retained control of the housing system, attempted to reassert authority over the way in which people were transformed into objects within the housing-class discourse. In November 1935 an office memorandum was issued, claiming that the system had become too complicated and that seniority alone should be used to allot housing.[11] Despite this, in 1939 accommodation was still given to individual departments that allotted the housing themselves. By this time, however, the system was coming under intense pressure from the shortcomings of the material forms that were necessary to support the emergence of New Delhi's specific class structure. This will be examined in the following section, after an examination of the lived nature of this class structure and a conceptual interpretation of the city.

Identity: Grids of specification

New Delhi marked the emergence of two complementary but different forms of hierarchy: one spatial and one social. They reinforced each other but were by no means a perfect fit. This can be explained by the disjunction between material and non-material elements of discourse. Yet, the complex negotiation of identity in New Delhi went beyond this. A 'grid of specification' emerged in the city that dictated how discursive objects would not just be classified and rated (Gutting, 1989: 234), but also divided, contrasted, related, regrouped and derived (Foucault, 1972: 42).

The physical 'grid of specification' took in not only location, but also bungalow size and design, plot size, street name and amenities, elevation and maintenance (King, 1976). For instance, one of the most prestigious 'bungalows', on King George Avenue, consisted of a hall, living room, dining room, study, sitting room, 6 bedrooms and bathrooms, 2 dressing rooms, 2 garages, 3 stables and 13 servants' quarters.[12] In contrast, the lowly peons dwelled in terraces of one room quarters.

The construction of the houses themselves was planned in accordance with the abstract surfaces of governmental finances as much as with regard to the local environment. The 10% of salary rent cap was used to estimate the cost of construction, although the Government had to subsidise the houses due to what had been stipulated to the PWD as New Delhi's special 'aesthetic and administrative necessities'.[13] The Gazetted Officers' bungalows were initially divided into categories A–D and varied substantially in size. The largest in class A totalled 201,864 cubic feet within a 4-acre compound, at a cost of Rs 62,403, while the largest in class D totalled 60,200 cubic feet in a 2½-acre compound, at a cost of Rs 27,262. However, per cubic foot, the class D bungalow cost 50 per cent more than the class A, yet the 10 per cent salary ceiling meant that the lower rungs of officers had to be subsidised to cover this cost. The clerk's quarters, however, received a disproportionately smaller subsidy, as became clear when the buildings were occupied.

While the PWD maintained the buildings themselves, the NDMC took over from the DTPC the more general landscape maintenance. Byelaws were created in an attempt to control all possible aspects of the city plan. These included stipulations from dog registration, hedge maintenance and bill posting, to where to walk in public gardens and how to act.[14] Most of the cases filed by the Committee concerned the landscape. For 1936–7 there were 283 cases of encroachment on public streets and 125 for unauthorised structures, with the next highest offence being just 19 cases regarding the provision of drains.[15]

Reinforcing the relational nature of housing in New Delhi, these grids of specification moved beyond the physical distribution and establishment of norms. Ordering also relied upon the relative placement of objects in relations of similitude and difference (Foucault, 1970: 67, from Steinberg, 2006). As such, Stoler (1995: 11) has argued that '… discourse on bourgeois selves was founded on what Foucault would call a particular "grid of intelligibility", a hierarchy of distinctions in perception and practice that conflated, substituted, and collapsed the categories of racial, class and sexual Others strategically and at different times'.

Beyond a simplistic gradational concept of class lies a relational, interactional sense of class as a cluster of practice that necessarily draws upon other aspects of identity formation (Duncan and Legg, 2004: 253). Racial logics did not draw upon ready-made class concepts. Rather, race, class, sexuality and gender were, and are, mutually constitutive elements of fractured and decentred identities (Stoler, 1995: 123). There was not a simple 'us and them' psychology based on race. Rather, the colonising elite were suspicious of, for instance, lower class white colonisers or threatening white sexualities (Levine, 2003). In addition, 'colonial racism' was not just about establishing difference. Rather, 'It was also how people identified the affinities they shared, how they defined themselves in contexts in which discrepant interests, ethnic and class differences, might otherwise weaken consensus' (Stoler, 2002: 25). Low (1996: 163) has directly related the colonial urge to suppress ambiguity, in favour of ordering and placement, to the spatial processes of the city. The urge to solidify fluid social categories that were constantly in danger of merging created an obsession with boundaries and segregation that was ever present in New Delhi.

The complexity of the grid on which people were placed in Delhi was made clear in a Finance Department file on allotment from 1938.[16] Applicants were listed in a table that detailed the coming together of class, geography and religion. The applicants' class was first denoted by their 'position', such as clerk or assistant (class by precedence). Their location was then separated into New or Old Delhi. Religion was assessed as being either orthodox or unorthodox, although this form of assessment was heavily mediated by considerations of culture (as discussed below). Class was further assessed through rupees earned per month (class by emoluments), which denoted the official 'class' category (from A to D). On consideration of these factors, an area was allotted in which the workers could live.

Yet, this classification is illustrative of most of the town planning discussions and the official *ethos* in Delhi in that it excludes considerations of gender. Women were increasingly valued by colonial administrations from the late nineteenth century (Stoler, 2002: 1). They were thought to guard racial vigilance and authority, and to prevent men from racially transgressive sexual affairs that came to be outlawed, or confined to the brothel. The arrival of women in the colonies contributed to the restructuring of colonial space. More strictly defined residential compounds and more elaborate housing environments sharpened the post-'Mutiny' segregation of European and Indian populations (Stoler, 1991: 225; Blunt, 2000). Women were provided with novels and books on conduct that antedated the forms of life they described (Stoler, 1995: 109; Blunt, 1999). These regarded not just dress and manners, but also the position of the house and separation of the servants' quarters from the home (Low, 1996: 161).

Such relations were present in New Delhi but obviously had their precedents. Nancy Dearmer recorded her trials adjusting to life in the Civil Lines north of Old Delhi as the wife of a teacher at St Stephens College in 1917.[17] She wrote of interrogating the servants to determine what their actual pay should be and organising the protection of the compound. This was not just from thieves but also from the 'box-*wallah*' salesmen who would descend on the veranda to display their wares. Dearmer's discipline over the marginal space of the veranda was matched by her time discipline throughout the day, interviewing the staff in the morning to give them orders, then dividing her hours between work, sewing and reading.

These spaces of negotiation for gender, class and race identities were extended into New Delhi as it was occupied. Viola Bayley wrote of her life in the new city as the wife of an employee of the Intelligence Bureau.[18] The city was divided upon the lines of class, and thus race, in terms of area. Yet, the more prestigious the house, the more Indian servants were needed to service it, and these people lived on site. As the manager of the compound, it was often the job of the wife to deal with these people. Bayley recalled making a periodic inspection of the servants' quarters in the summer of 1938. There she discovered not only the *dhobi* (washerman) in his quarters, but also his wife, mother-in-law, four children and several rabbit hutches. Speaking of a larger house near Kingsway that they occupied during the winter of 1938–9, Bayley detailed the tensions associated with negotiating the interaction between a high class, white woman and a low class, Indian man: 'A convention I always found embarrassing was the inspection of servants' quarters. The *mem-sahib* [Bayley] was expected to see that go-downs were clean and neat and the various wives and children reasonably flourishing. To me it always seemed an intrusion of privacy and unpleasantly patronising. Actually I doubt if the servants resented it and the women in purdah probably enjoyed it.' Bayley also spoke of overlooking the fiddling of the accounts in order to keep a good cook: 'It was rather like a ritual dance. Every move was known. Dignity was maintained but neither side must make a false step.'

Women's contributions were not, however, thought to extend beyond the home. In 1944 the War Department asked the Chief Commissioner whether British women in Delhi '... were pulling their weight in the war effort?'.[19] Chief Commissioner Askwith replied that his views would no doubt be considered heretical, but speaking candidly he felt that British women in India would be a liability in the workplace. There was no shortage of labour, Indian men were cheaper and it would have been a scandal to pay a woman a wage who was '... incapable of writing a coherent letter ...'. Askwith's condemnation was not all encompassing, he just conformed to the gendered, bourgeois geography that had collapsed in England, but could still be defended in India:

> The wife of a British civilian or regimental officer who discharges efficiently the functions which English women of the best type in India have always regarded as their own – looking after her husband and her children and doing something for the welfare of the men (both British and Indian) working under him – is performing services of the highest value. To conscribe such a woman to become an inefficient clerk in an army office would be stupid in the extreme.[20]

It would not be fair to portray the capital as a space that completely negated female agency. Women's powers in the home were considerable, and their involvement in social events and charity work allowed them influence at high levels of society. The Government's few female Indian employees could also exploit imperial self-conceptions regarding protectorship and duty. In March 1928, Miss I. Mitra, Assistant Superintendent for female education, was in need of housing.[21] She suggested the PWD build her a bungalow, but they refused and offered her an existing bungalow instead. The Education Office rejected the building as being unsafe for a single Indian lady. Mitra herself then wrote to the Registrar to the Chief Commissioner requesting a house with a few rooms, servants' quarters, and a garage nearby. She continued, 'The anxiety on my mind is most terrible. During the last fortnight it has told on my peace of mind and health. I don't want to break down just now. My work needs me.' This forced the Delhi Administration to break its own rule and pay Mitra's house rent while a suitable bungalow was constructed.

Religion also played a role in specifying the details of New Delhi's social grid, even if it was mediated through cultural categories. As mentioned above, Indian clerks' quarters were classified as either 'orthodox or unorthodox'. A series of petitions in 1915, to be discussed in the second section, had forced the government to consult its workers regarding the design of clerks' quarters. In 1917 it showed its redrawn designs to a representation of Indian workers, who demanded further changes.[22] They requested more bedrooms, private gardens and bathrooms. Yet, they also stressed that there was a certain proportion of Indian clerks who preferred to live in houses in the European style. It was requested that 5 per cent of the clerks' quarters in groups B–D be built in this style, and charged at European clerks' rates. The Government obviously thought this only likely, or desirable, for the better off Indian clerks, as class D quarters were not changed. Within a month orders were issued for 8 per cent of the new class B and 6 per cent of the new class C to be built in the European style, known as unorthodox quarters. Indian clerks' quarters came to be referred to as orthodox. This was not only a cultural term, denoting something in accordance with tradition, but also a religious term, as the orthodox quarters had walled courtyards in which female members of the clerks' families could sit in summer and still maintain *purdah*. The toilets and cooking facilities were also in accordance with cultural and religious traditions.

The further evolution of the un/orthodox balance reinforces the impression that the cultural and material, the discursive and 'non-discursive', cannot be thought of as separate. Through its aesthetics, lifestyles and hierarchical classifications, New Delhi changed the aspirations of those who came to live there. A representative body for the clerks, the Imperial Secretariat Association (ISA), campaigned for housing improvements in 1929.[23] The ISA representative stressed to the Government that D-type orthodox quarters were now viewed as totally inadequate for members of the Association: 'Further, the changing mode of living of the Indian members of my Association and their growing preference for unorthodox quarters, leads my Committee to recommend that if any new quarters are to be constructed, a large number should be of the existing "B" and "C" type unorthodox class.' By 1932 the Government had been forced to accept that a majority of higher ranked clerks wanted European housing, with the ratio of un/orthodox 'A' quarters standing at 55/18, while the ratio across all classes of quarters stood at 305 : 1,518.[24] Although still a minority, the unorthodox quarters now constituted 20 per cent of the total, a substantial increase over the 5 per cent requested in 1917.

This is an example of what Hacking (1986: 234) has referred to as 'dynamic nominalism': the way in which the categories that try to stabilise and order the world cause changes in the materials they classify. This admits the role of autonomous behaviour by the people who are labelled, destabilising the system with the resources it gave them. These are the enunciative modalities that will be examined in the second section, but the macro-changes in New Delhi's structure also represent this dynamism and the mutability of the colonial order of things.

The *Archaeology of Knowledge* has been criticised for failing to explain historical change (Gutting, 1989). Yet, while its general task is descriptive, it does carve space for sophisticated descriptions of historical change and memory, whether of concepts (Foucault, 1972: 56–7), transitions (1972: 166–77) or thresholds (1972: 186–9). However, Hannah (2006) is right to argue that we should not just focus on the 'surfaces' but also on the 'temporalities of emergence' of discursive objects. Indeed, Foucault's (1972: 74) insistence that systems of formations were not immobile, imposed from outside, or static, forces the consideration of systems of ordering as inherently mutable. The centrality of similitude, as well as difference, to hierarchies mean that it is the rank that remains the same, not the object: 'It is a perpetual movement in which individuals replace one another in a space marked off by aligned intervals' (Foucault, 1977: 147, from Steinberg, 2006).

This movement in New Delhi was annual. As the migratory staff moved to Simla, their applications were reviewed by the Estates Officer and their location reassessed. As people were promoted up through the ICS, so they were propelled through the city. From the European clerks' quarters, one could move to the junior officer's bungalows; a rising officer could hope to move westward beneath Kingsway to the heart of power, or towards a more spacious bungalow. While most landscapes provide a snapshot of a culture and society at one particular time, New Delhi represents much more of a cinematic landscape. This is not just in the theatricality of its 'sets' or the vast number of 'extras' who were used for the theatrical performances in the spectacular central buildings. Rather, the city was annually updated to provide constant footage of social change. In this sense, despite being rigidly structured by divisions that were overtly or covertly racialist, the city was spatially democratic. That is, when occupational social ranks changed, the city responded. Between 1911 and 1947, Indian political society underwent tumultuous change, and New Delhi had to represent this.

The Government of India Acts of 1919 and 1935 both increased the participation of Indians in their own self-government across the country. These contributed to, and reflected, the growing competence of Indians within the bureaucratic system. Increasing numbers of Indian men and women were being promoted to offices in the capital from the provinces, or working their way up the promotional ladder within Delhi. As the ranks in the social hierarchy became filled by non-Europeans, so New Delhi became gradually Indianised. The process was gradual, but can be illustrated by a comparison of seven of the most prestigious government departments during the seasons of 1929–30 and 1939–40.[25] On the eve of the city's inauguration in 1930, the Members and Secretaries, the two highest posts, of the Home, Intelligence, Finance, Legislative, Communications and Commerce Departments were all European, with only the Education, Health and Lands Department having an Indian Member, but a European Secretary. Indians occupied a few of the Under Secretary posts and were only equally represented at the level of Superintendent. In line with the system of diarchy introduced in 1935, whereby the administration of certain departments was devolved to Indian rule but the core departments maintained for colonial administration, there was greater Indian representation in certain departments by 1939–40. The Departments for Legislation, Commerce, and Education, Health and Lands were headed by Indian civil servants. While the Communication Department had an Indian Secretary, it was headed, like the key departments of Home, Intelligence and Finance, by European men, although across all departments there was greater Indian representation from the level of Deputy Secretary down.

This was expressed in the residential landscape of the city. The Indian Members lived at the prestigious locations of six King Edward Road, and one and five Queen Victoria Road, while the Indian Secretaries and Deputy Secretaries all took up residence south of the status-line of Kingsway. The mere fact that these civil servants were Indian indicates nothing, of course, of their political persuasion, but it does hint at the disintegration of the various conceptual race lines that had been geometrically strung through New Delhi's neo-classical layout.

The conceptual landscape

Having studied the emergence of objects, their ordering and mutability, Foucault (1972: 56) set about describing the concepts that organised these objects. Three rules were identified regarding the formation of concepts. They address forms of succession, co-existence and intervention. Forms of succession order series of statements and detail the dependence of statements upon each other. This is a perceptual process that mediates experience and classification, linking discursive objects together into series. De Montmorency's note of 1912 hinted at one conceptual form of succession, an order that could be adapted *mutatis mutandis* to different material sites of expression. This adaptation swarmed throughout New Delhi, creating a more general conception of social and spatial hierarchy. The archaeological concept informing this hierarchy was that of precedence. In this sense it constituted the *techne* of colonial urbanism, bridging the technological construction of the material landscape and the performance of the hierarchies it materialised.

The classifications of the Warrant of Precedence were felt throughout the social engagements that were attached to the Government of India, and were constantly adapted to different contexts. The blatant gradations of precedence were socially articulated, although not always appreciated. This was expressed by the Deputy Secretary of the Education, Health and Lands Department, and later Chairman of the Delhi Improvement Trust (DIT), Mr A.P. Hume (who will be discussed again in Chapter 4). On 27 June

1936, Hume wrote to his parents describing a *tamasha* (social function) that had been held in Simla to celebrate the King's birthday. To observe the parades and ceremonies he wrote that, 'we were all in our respective enclosures according the grade of mutton of goat. As a Deputy Secretary I was entitled to appear in gorgeous fancy dress of blue and gold, yet I preferred to use morning-coat.'[26] Hume commented more specifically on the spatial representation of the social order at a farewell ceremony for the Viceroy that was held at New Delhi Railway Station earlier in 1936. A semi-circular *shamiana* (an awning or canopy) had been erected to protect four rows of the 'middle and upper cut' from the sun. Hume described these as 'In official designation those who by the fortune of circumstance or office found themselves lower down the scale than article 23 of the order of precedence. Those superior to article 23 found favour beneath the fans on the platform.'[27]

The spatialisation of precedence occurred at all scales, from the micro to the macro, from the occasional to the everyday. Were one lucky enough to be invited to the Viceroy's House for dinner, the seating arrangement accurately mapped the Warrant's gradations. From this spatial grid one could thus select whom to talk to and dance with. As such, the city itself came to function as a social referent. Simply asking someone's address gave one an indication of his or her status and pay. The city also had a synecdochical relationship to the Warrant in certain situations, standing in for its complex gradations. During the inauguration of the city in 1931, detailed traffic regulations were issued to the participants.[28] While the Ruling Princes and Chiefs were given special parking and instructions, the residents of New Delhi were divided into those living north or south of Kingsway, and given directions to different car parks. This effectively divided the officers and higher ranks from the clerks and lower ranks through the use of a geographical reference. These groups were then subdivided into those allotted seats in blocks A–M or N–T. These blocks themselves had been organised into what Hume referred to as enclosures of mutton, further reinforcing the social hierarchy. It could be argued that the car parks were simply used to coordinate traffic flow, but the government often used such justifications. While de Montmorency insisted the city was designed to safeguard the greatest 'convenience' for each group, it also reinforced existing social, and racial hierarchies.

The historiographical analysis of this conceptual landscape has usually made reference to Indian traditions and their effects on the mindset of Raj officials. Hosagrahar (1992) has compared the layout of New Delhi to that of a Durbar. Here the Viceroy's House represents the Mughal Emperor, the British Viceroy or the King-Emperor. Around this point were organised concentric rings of status. However, this interpretation fails to acknowledge that New Delhi was not just colonial, and ordered around precedence. It was also modern and organised around conceptions of zoning. The division of the city into quarters for certain status group fractures the Durbar effect, although within the quarter, proximity to the core did increase status. Anthony King (1976: 63, 230) hinted at the commingling of British and Hindu conceptions of social hierarchy. Here the British posed themselves as superior to the caste hierarchy, yet simultaneously absorbed some of its logic and tendencies (Gilmour, 2005). The sub-groupings of New Delhi thus represent caste-like divisions of social space (Mitra, 1970: 46). This has often been acknowledged and was reflected upon at the time; Hume commented that he was received like a 'heaven born' (the Brahmin caste) at a State Reception, for instance.[29]

While offering some useful conceptual tools, such interpretations risk applying conventions of the nineteenth-century Raj to a twentieth-century capital. Following the 'Mutiny' there was great interest in Indian tradition as the means to access the supposed collective Indian mindset. This logic did enter the design of New Delhi, building on Delhi's reputation as a 'name to conjure with' (Viceroy Hardinge, 25 August 1911).[30] Yet, the city was also designed by British-based architects and reflected the new realities of the twentieth-century European metropole as well as the colonial periphery. The early twentieth century saw the rise of eugenicist thinking, which reframed traditional fears regarding cultural contagion. As the colonial state struggled to find new ways to express its authority, racial and class markers were strengthened to assuage fears about the collapse of social order (Stoler, 1991: 233–51). This was in line with a broader shift in colonial policy from assimilation to segregation. While affected by Victorian and Orientalist conceptions of hierarchy, the eugenicist obsession with boundaries and stock also frames the conceptual succession of discursive objects in New Delhi.

While Foucault's conception of the relation between discourse and material setting will be discussed in the following section, in addressing the formation of concepts he did consider synchronic and diachronic relations between 'co-existent' discourses (Foucault, 1972: 57–8). In terms of the former, a conceptual field of concomitance allowed statements from other discourses to be active within a simultaneous yet separate discourse. Old Delhi existed in a state of concomitance to New Delhi; the two cities co-existed with, and accompanied, each other. As such, objects from discourses that centred on the old city often became active in the new city, and the borderland between the two settlements often triggered this activation. Two of the most prominent discourses addressing the old city were those of policing and sanitation, both of which filled the empty space between the cities with meaning: as military *glacis* (cleared land around a fort that allows the effective use of gunfire) and *cordon sanitaire* (a gaurded line between infected and uninfected districts), respectively. Chapter 3 will show how the *glacis* was a product of post-1857 land clearances. It was maintained as both a practical and a symbolic space of distinction between the cities. The rise of anti-colonial nationalist protests in Old Delhi sparked off concerns in the new city that brought discussions about security and policing into the heart of the capital. Similarly, Chapter 4 will show that the *glacis* came to be conceived of as a *cordon sanitaire* as the congestion and disease of Old Delhi rose. Complaints were raised about the dumping of nightsoil between the cities, and the outbreak of cerebrospinal fever in New Delhi in the 1930s prompted Government action to help the older city.

In terms of diachronic co-existence, Foucault referred to fields of memory within discursive concepts. These were traces of discourses that were no longer accepted as true or as valid, but which still had relations of filiation and continuity with a contemporary discourse (Foucault, 1972: 58). In New Delhi the 'city' was haunted by the model of the cantonment. Settlements for Europeans in India had developed along the 'Civil Lines' model and always retained a degree of formalism and unconnectedness that interrupted the emotive sense of place that typifies city life. In planning the new capital there seemed to be a common assumption to begin with that it would be a glorified cantonment. Amongst the various submissions initially competing to influence the design of the city was a memorandum submitted by Mr L. Porter on 19 March 1912. This suggested that special legislation would be required, establishing the new 'lines' through '... a cantonment Act and Code worked by an autocratic commissioner instead of the present machinery of a cantonment authority and cantonment committee.'[31] Such ideas were obviously widespread enough to concern Captain George Swinton, Chairman of the DTPC. He stressed the necessity of beauty and dignity in: '... what I fear may develop into little more than a superlatively well arranged Cantonment ... .'[32] Similarly, one of Viceroy Hardinge's advisors wrote to him that 'It is not a Cantonment we have to lay out at Delhi, but an Imperial City.'[33] Colonial urbanism in the twentieth century more broadly sought to go beyond the military engineers' conceptions of urban space (Rabinow, 1984, 198). Yet, the failed development of New Delhi forced a reconsideration of the city's status in the late 1930s at which the memory of a more cantonment-like existence would resurface.

New Delhi also contained material memoryscapes of social orders whose influence was not what it once had been. The Cathedral in New Delhi had originally been allotted a much more prominent position, and a grander scale, than it finally achieved (Volwahsen, 2002: 263). Although there were some that thought of the capital as a soulless and godless environment, the religious establishment had actually secured its place within the socio-spatial hierarchy. The revised residential allotment rules for 1925 stressed that the Civil Surgeon and the Chaplain of the Church of England would be classed as 'B' officers irrespective of their pay.[34]

Even more prominently placed were the residences of the Maharajas. These were located around the Memorial Arch on Kingsway, which was itself a very different type of memory space. The Princes had their sovereignty assured by Queen Victoria's proclamation of 1858 and this was expressed in their elaborate dwellings (Volwahsen, 2002: 250). Their grand palaces were appropriately situated; they were at the heart of the Empire, yet outside of the swoop of the residential hierarchy from the peons in the northwest to the clerks and officers of central Kingsway, and the more prominent dwellings to the southwest. A note from 1931 listed the Princes alongside their status, as measured by the number of rounds of gunfire they merited on state occasions and the area of the site they had been allocated.[35] These ranged from the Maharaja of Limbdi (9 gunshots and 4 acres) to the Maharaja of Hyderabad (21 gunshots and 9 acres). The hierarchy of Princes was thus memorialised in the landscape, despite their circumscribed powers, not only in the face of the Raj but also of the emergent nationalist parties.

The final tool Foucault provides for analysing the conceptual landscape, beyond succession and co-existence, is that which examines procedures of intervention (Foucault, 1972: 59). Such procedures create new statements through rewriting, transcribing and translating information, or, through delimiting the transfer of statements. The unease created in New Delhi by the concomitance of sanitary and policing discourses can be seen as an intervention into the capital by discourses from the old city. Similarly, Chapters 3 and 4 will show how New Delhi intervened in the disciplinary and biopolitical landscapes of Old Delhi. Yet, the managers of the capital also sought to intervene to the north through transcribing their spatial hierarchies onto pre-existing urban forms.

The old city had not been totally neglected during the planning of the capital. De Montmorency had ordered the planning of a Western Extension for the city in March 1912, such that the DTPC could not later be accused of checking the commercial expansion of Delhi.[36] Viceroy Hardinge had even gone so far in April 1912 to suggest that the capital should comprise both new and old Delhi and that one officer should rule them.[37] This did not come to pass, with New Delhi being directly administered by the Government of India and the rest of Delhi Province being administered by a Chief Commissioner. Certain members of the Delhi Administration qualified for housing provision by the Government, but as the housing shortage developed in New Delhi it became obvious that they would not be accommodated in the new city. Rather, the Government appropriated accommodation outside of the city and slotted them into its residential hierarchy. Metcalf House, to the north-east of the old city, was converted into accommodation for members of the Indian Legislature.[38] A series of private bungalows were also rented to cater for class 'A' accommodation, but government-owned bungalows were used for 'B' class officers' accommodation and a range of clerks' quarters were arranged for 'A–C' classifications.[39]

Beyond this categorisation of space around Delhi, the capital was having other effects on its older neighbour to the north. The new city had become filled, and various workers had to live in Old Delhi and commute to the capital. The Chief Commissioner, J.P. Thompson, wrote to the Government of India on 21 May 1929 stressing that housing had to built for the clerks in New Delhi as congestion in the old city was pushing up the rates of tuberculosis (for more details of the emergence of this epidemiological register of congestion, see Chapter 4).[40] Little changed over the next 7 years and in March 1936 J.N.G. Johnson, the Chief Commissioner, wrote to the Home Department that he was left in the dark with regard to what development New Delhi would produce next.[41] This breakdown of communications was symptomatic not just of the Government's high-minded attitude with regard to the Delhi Administration, but also of a lack of coordination that threatened the intricately constructed socio-spatial hierarchy with collapse. Indeed, at the level of materiality and practice, New Delhi was a much less-ordered landscape than its files and imagery might suggest. It is from the production of colonial order to its consumption, from the crystallisation of hierarchies to their dissolution, that we must now turn.

The Spatial Dissolution of Order

Problematisation: Spaces and subjects of dissension

New Delhi was persistently problematised throughout its occupation. Even before 1926, its design and means of construction had been critiqued (see the first section in Chapter 3 for problematisation through crime and the second section in Chapter 4 for problematisation of the capital through health). Yet, when occupied, the city created possibilities for problematisation not through ideals, but through place-based voice and non-verbal material practice. These problematisations alter not just our conception of the city, but also our view of the archaeological methodology. After reviewing Foucault's views on subjectivity and the 'material', the policy of under-building in New Delhi will be investigated as the root of the dissolution of order in the city.

In shifting attention away from great thinkers to the profusion of discourse amongst its practitioners, Foucault seemed to be inaugurating a democratic moment for linguistic explanation. Yet, this potential was proscribed in the archaeological works through Foucault's desire to radically critique humanism, and thus his inability to refer to the intentions of individual subjects (Dreyfus and Rabinow, 1982: 60). This denies even the possibility of unreflexively shared practices in favour of rule-governed discourse. Foucault insisted, contra *Madness and Civilisation* (Foucault, 1967), that the object was not to reconstitute the experience of discourses, although he did insist that '... I have no wish at the outset to exclude any effort to uncover and free these "prediscursive" experiences from the tyranny of the text' (1972: 47).

So how *did* Foucault envisage the subject in *The Archaeology*? Subjectivity was tackled through the concept of 'enunciative modalities', which focused on the context in which people speak (Foucault, 1972: 50). The relevant questions, therefore, addressed who was speaking, the institutional sites from which speech was made, and the position of subjects in relation to discursive objects. These modalities tell us much about the form of speech, and shift the emphasis from some over-arching *langue* (language as an abstract system) to focus on the geography of *parole* (the practice and performance of language). Yet, what this system fails to explain is why some individuals come to occupy subject positions that others do not (McNay, 1994: 76). How do individuals decide what to do with the position they occupy?

Foucault does offer some hints as to how a more complex archaeology of the subject could be worked out. The book is peppered with references to the creation of possibilities, choices, heterogeneity and, especially, contradictions. The last were not to be glossed over, but to be described as a means of revealing different or similar ways of approaching the world, of localising the divergence and juxtaposition of discourses. The task was to describe different 'spaces of dissension', to analyse the different types of contradiction through mapping their levels and functions (Foucault, 1972: 152–3).

Two types of contradiction were identified. Derived contradictions are localised assertions that do not affect the body of enunciative rules and can originate within the same discursive formation. Intrinsic contradictions are deployed at the level of the discursive formation itself and can emerge between discourses. These contradictions not only play out across the levels of discursive formations, from objects to themes, but they also have different functions. They can lead to the additional development of the enunciative field; starting experiments and making new objects possible. They can also transform the discursive field, translating statements to different contexts that re-forge them as new discursive objects. Finally, contradictions can play a critical role: 'they put into operation the existence of the "acceptability" of the discursive practice; they define the point of its effective impossibility and of its historical reflexion [*sic*] ...' (Foucault, 1972: 155). Discourse as such is not smooth or resolved: 'It is rather a space of multiple dissensions; a set of different oppositions whose levels and roles must be described' (Foucault, 1972: 155).

Contradictions, thus, open up greater spaces of choice through which we can glimpse individual agency. Foucault would respond to the criticisms of anti-humanism in his later governmental work on the 'conduct of conduct' (Foucault, 1978a [2001]), counter-conduct (Foucault, 1978b 2007), the *Use of Pleasure* (Foucault, 1986a) and the *Care of the Self* (Foucault, 1986b).

It is with the retroactive faith of these later texts that one can insist that although a subject category is derivative, it may also be effective (Hannah, 2000: 41). Indeed, feminist and post-colonial scholars have stressed that we cannot bypass the subject or assume its dissolution because what this actually leaves us with is a world that privileges white, male thought (McNay, 1994: 79). If all subjects are fragmented, is there no space from which to construct alternative identities? Spivak has famously suggested that we need to strategically essentialise subject positions to allow political identities to emerge (see Spivak et al., 1996). This has been taken up in much post-colonial thought as a call to discover the subjective voice of the colonised as against the objectified discourse of the coloniser. Yet, this is antithetical to Foucault's (1979b) suggestion that repression leads to a multiplicity of discourses (Young, 2001: 407). Rather, discourses create multiple enunciative modalities, the contradictions between which can allow these positions to be used critically and in resourceful ways.

The discussion of New Delhi's grids of specification showed that material space could be analysed as contributing to and reinforcing the striations of discursive space. Yet, when Foucault wrote directly about material space in *The Archaeology*, the relationship between written, embodied and physical space was not as clear as it had been in previous works (see Philo, 2000). He did insist that discourse was not about signs or language, but that it referred to practices that form the objects of which they speak (Foucault, 1972: 49). Similarly, he insisted on the relationship between discursive and 'non-discursive' domains, such as economic practices and processes, institutions or political events (Foucault, 1972: 157, 162). However, the articulation of these two realms was explained as the appropriation of the non-discursive by the discursive (Dreyfus and Rabinow, 1982: 77). Behavioural patterns, systems of norms, techniques or types of classification are said not to define a discursive object. They simply enable it to appear; they are its field of exteriority (Foucault, 1972: 45). Rather than having affect, space, even for Foucault here it seems, is dead (contra Foucault, 1980).

Yet, Foucault went on to become a great proponent of the affectivity of space. Viewed through this lens, his discussion of the non-discursive seems peculiar, as he later argued effectively that discourses were as much constituted by materiality and performance than knowledge or language (Laclau and Mouffe, 1985: 108; McNay, 1994: 70). As such, Dean (1994: 17) can argue with retroactive faith that archaeology is a materialist approach to the analysis of knowledge. Similarly, Young (2001: 399–400) argues that an archaeological analysis of discourse is not linguistic but addresses the materiality of language. McNay (1994: 79), however, remains convinced that, at this stage, Foucault's inability to fully conceive of the non-discursive left him with a rigid taxonomy that says little about the functioning of discourse in socio-historical context. This chapter confronts this rigidity through a study of enunciative modalities that stresses the contradictions that fissured these subjectivities, and that were mobilised to problematise governmental regimes. Second, the material realm is considered not just as a space in which discursive practices were played out. Rather, the material is here also viewed as governmental excess. Material spaces, and the lives that played out in them, confronted the government's vision of how New Delhi should appear. In this sense, the material spaces of the city resisted the government itself (see Joyce, 2003). Just as power and resistance cannot be separated, so the resistance of the material realm here owes its origins to the attempts made by the government to impose the Warrant of Precedence onto the capital's landscape without sufficient financial investment. This was a pre-justified policy of 'under-building' that sets the context for the subjective critiques of the contradictions of governmental housing policy.

While the Government of India meticulously planned its conception of social order, it did not devote as much attention to tracking the development of this order on the ground. From the beginning of the capital planning project, there was uncertainty over how many people had to be accommodated. As such, warnings of the inadequacy of the proposals went unheeded and plans to not provide housing for all inhabitants were rubber-stamped.

From the very beginning, Viceroy Hardinge insisted that limits be put on the number of workers to be housed by the Government. On 12 February 1913, he dismissed Lutyens's estimate that the city would have to occupy 5,630 workers.[42] He claimed that 2,500 of these were press-hands that did not have to be placed in the city (although they later were) so he reduced the estimate to 3,000. On the same day, Hardinge wrote to Sir Malcolm Hailey, the new Chief Commissioner. He pointed out that many officials could not afford to live in separate bungalows and that cheaper flats would be required.[43] On 3 June, Hailey ordered an estimate to be made of the number of people to be accommodated. These estimates were obviously not to the IDC's liking as by the end of the year it had decided that it would be 'unnecessary and extravagant' to provide for all temporary and permanent officers and subordinates. Its report from 29 December 1913 stated that private enterprise would be encouraged to provide accommo-dation; '... provided always (and the Committee wish to strongly emphasise this proviso) that Government is prepared to build further accommodation should a tendency be developed to force up rents.'[44] The different classes and the percentage that would be unprovided for were listed as follows: officer I (20 per cent), officer II (40 per cent), officer III (60 per cent), officer IV (70 per cent), married European clerks (15 per cent), unmar-ried European clerks (85 per cent), Indian clerks (10 per cent), menials (25 per cent). Hardinge's suggestion that lower class officers could not afford the bungalows was taken on board, although the clerks were comparatively well provided for.

These plans immediately created unease. The Secretary of the Home Department suggested on 19 January 1914 that the unprovided portion was too large and that private enterprise would not fill the gap.[45] The PWD not only responded that hotels, clubs and chambers would absorb some of the excess, but also stressed that even private enterprise would charge rent at 10 per cent of the outlay cost, which would be ruinous to younger officers. The Secretary of the PWD, R.P. Russel, also cast further doubts on 3 March 1914. Through looking at the land allotted for private Indian and European housing, he could not explain how the excess government officials would be housed.

Despite Russel's objections, on 17 March 1914, an Order in Council set the official under-building rate for officers' bungalows at 33 per cent.[46] By September 1914 the PWD could report to the India Office in London that only 67 per cent of officers' residence and 80 per cent of clerks' residence would be constructed. This was in the context of the (First World) War that had been declared in August 1914, causing financial cutbacks across the Empire. By August 1915 the PWD had worked with the IDC to incorporate the limits in expenditure and produce a new housing scheme.[47] The number to be provided for had been reduced by 85 officers, in the belief that only the camp army, not the whole Army Department, would be located in the capital (although they later were). Of the 29 officers associated with the Chief Commissioner, only 12 were given accommodation. From the total of permanent resident staff, under-building deductions of a third were then made, leaving 154 houses for 246 officers. However, with regard to officer class I, the PWD stated that 'Since any reduction in class is unlikely, a corresponding reduction from class VI has been made.' This ensured that the top officers would be well provided for, but that the lesser officers would be left to a private market that the PWD itself had described as ruinous.

This plan began to take shape over the next few years. In December 1916 the under-building rates were confirmed as officer I (zero per cent), officer II (35 per cent), officer III (43 per cent), officer IV (56 per cent), married European clerks (28 per cent), single European clerks (40 per cent) and Indian clerks (44 per cent).[48] However, through adjusting the estimates of how many temporary staff would have to be given accommodation, the actual number of officers and clerks to be provided for had dropped from the original estimate of 3,000 (down from 5,630) to just 1,147.

By 1922, concern was growing amongst the government at the cost and expedience of the capital's construction.[49] A 'New Capital Enquiry Committee' was established which cut the funding from Rs 1,307 lakhs (Rs 130,700,000) to Rs 1,292 lakhs. This also showed that most of the housing was either complete or under construction and that provision had risen slightly from the 1916 estimate to 1,289. Yet, the report was one of the first to indicate that the under-building policy was going to lead to major problems for the capital. The planned provision for migratory personnel was 136 officers' bungalows and 641 clerks' quarters. Yet, the demand in 1921 was for 254 bungalows and 1,140 quarters, indicating an under-building of 46 and 43 per cent, respectively. Plans to keep fewer men permanently in Simla would increase these figures to 54 and 45 per cent, respectively. The committee recommended increasing the officers' residences by 40 units, bringing under-building to 41 per cent, and the clerks' quarters by 347, bringing under-building to 15 per cent.[50]

By 1925 it was obvious that the number of menial quarters was also proving problematic. Office workers and record sorters had requested 210 married quarters and the New Capital Enquiry Committee had pointed out that the 100 provided were insufficient.[51] However, it was decided to build just 50 more in the hope that others would not need their accommodation for some time. It was this lack of precision or foresight that lead to the rapid acceleration of complaints and housing crises after the New Delhi offices were fully occupied during the winter season of 1926–7.

The Home Department immediately complained that while the unorthodox clerks' quarters were under-built at just 13 per cent, orthodox clerks' quarters were leaving 46 per cent without housing.[52] Tensions would grow over the following year, and on 19 March 1928, Mr Gaya Prasad Singh barraged the Government in the Legislative Assembly with 10 questions regarding accommodation in New Delhi.[53] The Government admitted that under-building had resulted in 21 per cent of gazetted officers being without accommodation. For unorthodox clerks the figure was 30 per cent, whereas 53 per cent of orthodox clerks were not provided for, despite the revised aim to cap the under-building for the Indian population at 44 per cent. It was becoming increasingly obvious by late 1928 that something had gone seriously wrong with the planning of the accommodation.

The Secretary of the (Old) Delhi Municipality provided a succinct description of the problem in a note of 7 September 1928.[54] It was pointed out that the Government actually employed 3,500 assistants in total, not the 1,147 the plans had been based on, and only provided accommodation for 1,500. The Railway Offices had recently transferred to the city, bringing 1,400 clerks, while the Post Office Headquarters was in the process of transferring. As the construction of the new capital drew to a close, private enterprise was growing, while many of the contracted labourers were staying on. The latter constituted 12,000 workers who had been brought to Delhi to construct the 1911 Coronation Durbar and retained to build the capital, but were then 'turned adrift' without receiving their full pay.[55]

A.M. Rouse, who had expressed his doubts about the accommodation plans in 1927, drew up the PWD's appraisal of the situation in March 1929.[56] Rouse had shown that there was a shortage of 42 officers' residence that he planned to remedy by building 12 A and B class bungalows and adding 30 hostel rooms for lower officers. He did, however, note that this provision would only bring the houses up to the 33 per cent under-building standard, whereas private builders had failed to provide this extra third of housing provision. For the clerks, in 1928–9 the government had only housed 1,680 of the 3,851 in need. It was claimed that there was demand for 2,383 quarters, which at 65 per cent to take under-building into account, left a demand of 1,891 against the supply of 1,541. As such, 350 more quarters were needed to meet the under-building standard. Rouse concluded that, 'One thing, however, is absolutely certain and that is more accommodation must be found, and that quickly'. It was in response to this report that Chief Commissioner Thompson had added his support, on 21 May 1929, to the urgency of Rouse's report with his warning about tuberculosis in Old Delhi as a result of over-crowding.

Under this pressure from the experts regarding New Delhi's infrastructure and Old Delhi's human geography, the Department of Industry and Labour responded on 1 June 1929 with an incredibly rare climb-down. In May 1929, an internal review of housing policy had concluded that under-building was too high, private provision of housing was too low and that the transfer of the Army Department would significantly worsen the situation. The Department of Industry and Labour had appointed a committee to investigate the problem. They concluded that the government needed quarters for an extra 103 officers and 858 clerks, at a cost of Rs 120 lakhs. The shortage was blamed on four failures of calculation. The 1923 building scheme had failed to take into account not only non-migratory staff, but also the head-quarters that had not yet been transferred to Delhi, the increased demand for personnel during the 1920s or the possibility that the 35 per cent under-building rate was too high. An under-building rate of 10–15 per cent was, as such, recommended, although this was revised to 22 per cent for officers and 14 per cent for clerks. However, more excuses were found not to engage in large-scale construction. The Simon Commission was, at that time, undertaking its constitutional review. Since this would affect the number of people working in the capital, any action was delayed.

The Chief Commissioner immediately set about compiling more evidence to force the Government's hand. In particular, Thompson wanted to show where the Government's workers were actually living. Since he was using alternative sources of information to official Government figures, no citywide total could be produced with absolute accuracy. Yet, the ISA was able to show that only 159 of the Railway Department's workers lived in New Delhi, compared to 577 in the Delhi Municipality outside of the walled city, 2,205 in Old Delhi and 88 in the Civil Lines. The Government also came under continued attack from the Legislative Assembly. On 23 February 1929, the Government was forced to admit that New Delhi could not house all its workers, but refused a conveyance allowance to those forced to live in Old Delhi.[57] On 9 March 1931, it had to publicly admit that for B class clerks' quarters, 58 and 48 per cent of orthodox and unorthodox clerks, respectively, were unprovided for.[58]

The latter admission was made in public just a month after the city had been inaugurated. Intended as a showcase of the might of British imperialism, New Delhi had been unable to house the people required to run the capital. The project had created new discursive objects and organised them on a meticulous grid, which fed into concepts of socio-spatial hierarchy that infused the city's functioning. Yet, these discursive formations were attempting to operate in a position of autonomy with regard to the material realm, embossing at will their hierarchies onto the urban form. Like Foucault's vision in *The Archaeology*, such discursive simplicity was complicated in the material world of practice and existence. A perfect material rendition of colonial hierarchies was thwarted by the Government's own lack of calculation and its inability to extend its vision over the seething multiplicity of the material world it had created. Yet, besides people's very existence, there was an ongoing critique of the capital within. This was not so much a finding of voice by the subaltern, but a critique that was created by the capital and operated from spaces that the city actually sustained.

Identities: Enunciative modalities

Petitions

To assess the subject positions of a discursive formation, Foucault (1972: 50) suggested that we first ask who can speak, what level of competence is required, how speakers are divided and how this speaking relates to others. In New Delhi one's ability to speak to the government was dictated by one's rank. Since each government employee had the right to apply for accommodation, it also gave each employee the right to address this right. The superior and subordinate members of the staff, through both formal and informal channels, used petitioning to challenge the hierarchy of accommodation in New Delhi (see the first section in Chapter 3 and the second section in Chapter 4 for further petitions requesting governmental assistance). These channels were also taken up by non-governmental employees, although their ability to exploit these avenues of complaint were limited. Petitions were used not only to question the structure of the city, but also to question its *derived* contradictions, those everyday problems of existence that should have been catered for. These separate petitions, though not formally linked, can be seen to form a counter-discourse that is unified by its opposition to the accommodation scheme of the Government.

For instance, a petition was submitted to the Government in 1916, signed with fingerprints and written in Urdu, but translated into English.[59] The petitioners had been ejected from their original inhabitance in June 1913 when it was acquired for the capital construction. They had since been living in a private garden, which they had been assured would not be acquired. When this took place in March 1916 they petitioned for compensation. The petition was unsuccessful, with notes on file suggesting that the people could simply move their houses and that, although they had received bad luck, technically it was their own fault.

Similarly, in March 1927, the land and houses surrounding a small mosque in the New Delhi area were acquired.[60] The residents petitioned for compensation and for permission to take their housing materials with them when they were evicted, as the cost of such materials had been greatly inflated due to the capital construction. While compensation was paid, the material was not granted them as, the Superintendent of Monuments claimed, it was 'objectionable to show charity at the expense of government ...'; the materials would have to be bought back at auction. Both of these cases targeted small-scale failings of the government's aim to provide evictees with the resources to relocate, but failed due to the legal and technical machinery at the Government's disposal.

While these people could speak, they were not speaking from positions of strength. In looking at enunciative modalities, Foucault (1972: 52) stressed the importance of subject position. This encompassed not just the relationship of subject to object, here of the petitioner to the housing or land, but also the channels by which speech was facilitated. Those people employed by the Government, rather than relocated by it, had stronger claims to be heard in their petitions.

On 2 January 1918, the clerks of the Military Department submitted a petition for house rent allowance.[61] The clerks had not been given housing near the Secretariat, which at that time was located in its temporary building to the north of Old Delhi, but were living in the few completed quarters in New Delhi. They complained that tonga expenses were too high, and the 7-mile trek made it impossible for them to start work at 8 AM. The Raisina quarters were also said to be solitary and prone to theft and harassment, although this criticism would continue for decades (see the first section in Chapter 3). As such, in February 1918, the Home Department acquiesced and granted compensation.

Clerks could also adopt more aggressive stances. In February 1925, a number of clerks were accused of subletting their quarters.[62] Mr Ghulam Hussain replied that he had not sublet to anyone but that he had exchanged quarter with a friend 'for mutual convenience', and that the maximum rent was still being paid. He even suggested that no rules against subletting existed in Delhi, and that 'doubling up' was positively encouraged. Yet, while de Montmorency had tried to claim in 1912 that the city was a structure designed upon the lines of greatest convenience, the actual emphasis in the city was on people knowing their place. An unnamed official wrote on the file: 'We can't have this sort of thing. It makes trouble for everybody. The rules must be observed.'

Yet, those in more senior subject positions had different channels for petitioning open to them. For instance, the Education Commissioner with the Government of India wrote directly to the Chief Commissioner in May 1927 claiming that his superintendent was forced to live a great distance from Raisina, and that he would appreciate a nearer house being arranged.[63] As New Delhi was more fully occupied from 1927 onwards, complaints from senior figures about the failure to provide housing started to flood in. Unlike the clerks, these protests did not have to take the form of petitions in order to be processed by the Government. Reports were logged from two members of the Army Department who had to pay to send their families back to Bengal as no quarter was available to house them in.[64] Other complaints record men having to share apartments, or to leave their families in Simla rather than bring them down to the winter capital.

While these derived contradictions between the Government's claims and reality would continue to be picked up throughout the colonial period, petitions were also used to target more fundamental, *intrinsic* contradictions. These did not address the way in which the city was realised, but the hierarchies it sought to impose. The case of a memorial submitted in 1915 highlights not only the willingness of low-ranking government employees to criticise the intrinsic *ethos* of the city, but also that the Government had provided the opportunity for clerks to take up this position. On 2 September 1915, R.P. Russel explained that he had given clerks the opportunity to see the designs of the quarters that were going to be built for them.[65] Russel proposed to form a committee of clerks to feed back their ideas to the PWD, yet before he had the chance, a memorial had been submitted to the Viceroy himself. As the Home Department, who conveyed the memorial, commented of its own clerks:

> ... they have already submitted a memorial in which they have taken exception to the principle of differentiation in respect of the capital cost of the quarters intended for Anglo-Indian and Indian clerks and to the classification adopted in the case of quarters for the latter.[66]

It was this objection to the principles of differentiation, and classification, that marks out the intrinsic contradiction being targeted here. The memorial begged forgiveness for troubling His Excellency in a time of war, but insisted that the matter effected the memorialist vitally, and the status of clerks generally. The initial proposals had made a distinction between the capital cost of quarters for Indians and Anglo-Indian (white) clerks.[67] For instance, the top-grade married quarters for Anglo-Indians would cost Rs 9,200 to build, but only Rs 4,072 for Indians, which the memorialist showed was lower than for the cheapest Anglo-Indian quarter. The Indian estimates had been based on houses in the temporary capital, although it was claimed that these themselves were insufficient. But in summing up, the memorialist moved back from derived to intrinsic contradictions:

(2) There being no difference in status, pay and allowances, between Indian and Anglo-Indian clerks of the Secretariat, it would, it is submitted, be invidious to draw a distinction in respect of quarters ...

(3) The proposal amounts in effect to building better quarters for Anglo-Indian clerks at the expense of Indian clerks ...

(5) This unequal treatment is felt as a mark of inferiority of the Indians as a class.

(6) The argument, which is sometimes used, that Indian clerks generally live more economically as regards house accommodation than Anglo-Indians, fails to recognise the fact that such economy is in most cases a matter of compulsion.[68]

The memorial drew out a clear use of racial categorisation, showed that it was insulting, and then showed that the essentialising stereotype that it inherently deployed, of the humble Indian, was an imposition, not a cultural condition. The Home Department stated on 6 September that it planned to cooperate with Russel in organising representative committees for the clerks. On 11 October, the PWD confirmed that a committee had been established for each different housing class, and that they would be consulted as designs were produced.[69] They were given the chance to comment in February 1917 and some of their recommendations were taken on board.[70] These generally involved adding more rooms, but the clerks also insisted upon housing being provided in the 'European style', as discussed earlier in the chapter. This reinstalled a division of housing for people of Indian origin. Yet in this case, the division was instigated from below and was a matter of choice, not racial designation.

As in the case of the derived contradictions, the petitioners seemed to become more vociferous in their criticisms over time. In 1937 the Chief Commissioner complained that servants' quarters in New Delhi were being rented out, forcing servants to live in garages, and making a profit for the home dwellers at the Government's expense.[71] The Government complained to Mr Lachman Das Bhandari of this, who replied on 12 September 1937 that this activity was not forbidden in the lease. The plot was still being used for residential purposes and, as such, Bhandari stated that: '... it is my own lookout as to who should live in different rooms or parts of the house.' He went on to ridicule the Government for suggesting they could dictate what one could or could not do in the private spaces of ones home: 'I daresay such a law is unheard of and does not exist anywhere in the world ... To restrict this discretion is to interfere with personal liberty.' He also pointed out that people had actually been building extra tenancies in their plots in the more elite areas of the city, yet there seemed to be a different law for those people. However, a city engineer pointed out on 25 November that the lease dictated that each building had to serve the function allotted it in the plan, and that paying tenants could not be designated guests. While falling down, again, on the legal technicality of the lease, Bhandari had picked up on a key contradiction of the British 'liberal' Empire and had reclaimed one of the key organising concepts of the imperial landscape. Namely, that it claimed to respect the liberty of the individual yet intervened into the lives of its subjects to an extent that would not have been tolerated in the West.

Sites of authority

While Foucault drew attention to the subject's position, he also stressed their institutional site (Foucault, 1972: 51). These sites describe the places at which discourses are produced and consumed. The counter-discourses issued from the site of the home, while the success of the petition depended upon the subject position of the dweller. Yet, there were also representative bodies that took up both the derived and intrinsic contradictions of the Government to greater effect than an individual, irrespective of his or her subject position, could do. The two institutional sites that will be studied here were not marginal, occupying the space of the exiled or subaltern. Rather, they were central, occupying spaces created by the Government to fulfil its image as a liberal machine. The Legislative Assembly had not featured in the original plan for the city, as the right to wide-scale democratic representation was only granted to the Indian people after the Government of India Act (1919). The building was placed to the north-east of the North Secretariat, proving as problematic to the symmetry of the city as it had done to the authoritarian aspirations of the government. The second site was more central still. It did not have a permanent physical space but, rather, occupied the Secretariats themselves. The Imperial Secretariat Association (ISA) represented the views and grievances of all those who worked for the central Government of India, having 657 members in 1931. The ISA was a relatively conservative body, but in the liberal tradition it was a constant complainant against the Government, and did occasionally slip into more radical language. The Association had been producing a bulletin for private circulation ever since occupying the New Delhi Secretariat in 1927. In 1929 an editorial complained that the Government only ever offered compensation to those who did not need it, or could not claim it. It was concluded that such a situation could not be remedied under the present system: 'And so, it will have to be relegated to the category of dreams which we trust will come true when we have Dominion Status.'[72]

Since the Government of India centrally administered New Delhi, its day-to-day affairs could be directly raised in the Legislative Assembly. As such, the records of the Assembly capture a host of seemingly banal questions regarding the *derived* contradictions of a government unable to deliver on its housing of 66 per cent of the working population, or to keep the residences in a fit state of habitation. The earliest criticisms regarded *amenities*. Complaints targeted the lack of street lighting in the centre of New Delhi (1921),[73] which the Government stated it could not afford; leaking apartments (1923),[74] which the Government claimed were still habitable; unsafe housing (1927), to which the Government did not respond;[75] and parts of the city that were unsanitary (1928).[76] The last was with regard to a dumping ground that existed next to some E class clerks' quarters that had caused an excess of flies, although the Government insisted it had recently removed the dump and landscaped the area. Mr Lalchand Navalrai asked in 1934, 'in the interests of the smaller people', whether they would be graced with the flush latrines being installed in the officers' bungalows, but no plans had been formulated.[77]

In 1929, the ISA started to add to these calls, providing lists of the complaints its members had raised. When migrant workers returned to the capital from Simla, flats were often dirty, furniture was missing, broken or inadequate, while the buildings themselves were in need of maintenance.[78] A series of recommended changes were submitted to the Government, having much in common with those of the 1917 committees, although little action was taken.

Following on from these deficiencies in amenities, a second strand of critique addressed *compensation*. The context for such requests was the popular knowledge regarding the cost of New Delhi. In 1921 the Government was forced to admit in the Legislative Assembly that the incomplete city had already cost Rs 490 lakhs (between £3 and 4 million).[79] As such, demands were made that bus conveyance be provided for clerks living far from work (1925), while the ISA insisted that conveyance allowance be continued into 1927 for those 'who have to live in Old Delhi owing to the paucity of quarters in New Delhi'.[80] The Government's admission to the Legislative Assembly that it had cut the conveyance charge in 1929 because 'the city is within comparatively easy reach of the New Secretariat' was met with incredulity and a question to the Industry and Labour Member as to whether he could walk the distance and attend his office in time?[81] There was no reply.

At times the criticisms of the ISA and Legislative Assembly moved beyond the derived contradictions of a bureaucracy that could not fulfil its pledge to house its workers in suitable accommodation. At the *intrinsic* level the discursive formation of the landscape itself was attacked. On 22 February 1921, Khan Sahib M. Ikramullah Khan asked the question quoted below as the first of five points criticising the *distribution* and *allocation* of governmental quarters:

> Is it a fact that racial distinction has been observed in the construction and allotment of quarters in Raisina for the Secretariat assistants and clerks? If so, will Government be pleased to take early steps to remove such distinctions?[82]

Yet, the racial distinction he alluded to was that between Indian and European style quarters, a distinction that had been enforced by Indian clerks themselves. The issue of race, however, was obviously one that was stimulated by a city so obviously divided upon racial lines, even if one could choose to live in the European style. Given the clear topography not just of race but also of class in the city, the question of allocation was obviously a vexed one. In September 1925, a question was asked in the Assembly regarding the 'criterion or canon' by which quarters were allotted to members of the Assembly itself.[83] The utilitarian response was that '... the principle that directs is that of securing the greatest convenience of the greatest number', falling back on the 1913 premise of convenience when the city was proving convenient only for a very privileged minority. Further questions addressed the calculation of rent (1927), to which the government explained the system of 6 per cent of capital outlay but not more than 10 per cent of rent principle.[84] In 1931 a question was posed regarding the allotment of clerks' quarters, to which the government refused to spend the labour collecting the information for such a 'small benefit'.[85] A further question pushed for more information regarding the actual system of allocation, claiming that clerks suspected that allocation was not made in strict accordance with the waiting list, 'which is said to exist but which no assistant or clerk who is affected has ever seen'.[86] The Government insisted that the relevant Departments kept waiting lists. These questions pushed at the mechanism by which the regular administration of the uneven resources in the city was orchestrated. However, the Government's response to the immediate over-crowding of the late 1920s threatened to bring to light the most intrinsic contradictions of the city itself.

It has been shown that by the mid-1920s the Government had realised that there was going to be a housing shortage, but that it did not respond by building substantially more houses. While it claimed to have enforced a lower under-building rate for the, mostly Indian, clerks, the archival records actually reveal a *redistribution* of housing stock from the lower ranks to the, mostly British, officer class. The Chief Commissioner hinted at such a policy on 17 November 1926 when he suggested, 'Now that the bungalows are being re-arranged in Old and New Delhi it may be possible to induce Government to add a few more residences owned by Government to the few that exist at present for the use of officers.'[87]

The Chief Commissioner did not give details of his plans, but 3 years later it became apparent that a similar scheme had been put into operation. On 24 January 1929, the ISA wrote to the Secretary of the Department of

Industries and Labour, PWD, to address issues of housing shortage, classification and allotment.[88] The Association stressed that only 62 per cent of unorthodox and 50 per cent of orthodox clerks had been accommodated, and that this shortage had been accentuated by a new housing classification scheme. This reclassification of quarters according to pay had raised the emoluments necessary to gain access to an A class clerks' quarter, thus excluding many senior assistants from the top-grade quarters. The ISA denounced this as a '... harsh principle and an undesirable practice. This process, if continued, will gradually tend to relegate the Secretariat establishment to lower and lower class of quarters by the simple device of changing the classification'. The letter also stressed that because of the changing mode of living of its members many more clerks preferred unorthodox dwelling and very few found D quarters adequate or suitable. With the immanent arrival of the Army Department in the capital, a worsened situation was expected and the restitution of the previous classificatory system was requested.

This did not happen, however, and the reason for the stockpiling of quarters in the top bracket became clear over the following few years. The ISA had obviously grown tired of its requests and warnings going unheeded since 1927 and wrote in frustration to the Secretary of the Home Department on 17 December 1930 that '... the object of the formation of the Association is entirely frustrated if the specific requests made by them are ignored as in the present interest ...'[89] They included a letter that had been sent to the Superintending Engineer of the Central PWD. What the letter reveals is that housing had been stockpiled in clerk category A such that it could be converted into officer class E houses. This was the class that was identified in 1914 as being susceptible to the 'ruinous' private market, although the clerks had to face the same market but with much lesser pay. The letter stressed that this system would augment the grievances of its members, dislodging men who earned Rs 501–600 from A to B class quarters, where shortage was already greatest. The protest was not just at the derived contradiction of people not being accommodated in comfort, but it was also at the intrinsic contradiction of a government that enforced its own rules in an uneven manner:

> The members of my Association feel very strongly in this matter of such inequitable treatment and lowering of the standard of living. In this connection, I am to emphasize the fact that the principle underlying the Fundamental Rules that Government should provide accommodation appropriate to the status of the officers concerned would be infringed by the present proposal as two men drawing the same emolument would be allotted different classes of accommodation.

The ISA levelled its suspicions at members of the Army Department, who at the time only qualified for B or C class clerks' quarters, not officer's housing.

The rules governing housing allocation dictated that if an assistant could not get accommodation in their emolument class they would be offered the first class above, then the first below, then the second above, etc.[90] The surplus of houses in officer class E, and the deficit of clerks' quarters, would allow the Army Department to barter its workers into the recently converted houses, not quarters. The Association stressed that this would lose the Government money, as rent was fixed at 10 per cent of salary, and these men would be earning salaries equivalent to clerks class B or C. In sum, the ISA felt that

> If more quarters are needed, then build them, do not reclassify ... Instead of progressive building, the process so far has been one of reclassifying and raising the pay-limits thus lowering the standards of living contrary to admitted principles.[91]

The Government, however, did not seem abashed by this penetrative analysis. On the contrary, it continued to seek out ways to alter the classifications so that fewer people could compete for the elite housing of the officers. In 1933 the rules for allotment of officer's residences were amended to redefine emoluments.[92] These would no longer include benefits, allowances or pensions, so that people would technically earn less. This served to reduce demand for the more elite houses and repressed the claims of the increasingly Indianised lower ranks. In 1939 the ISA was still petitioning the Government to alleviate the hardship caused by its system of housing allocation.[93] By this point, however, the material environment of the capital had become so disorganised that a wholesale re-visioning of the capital was underway.

Lived space

While this section has examined the identities and subject positions that were created by, and mobilised in, the new capital, it has privileged speech over sheer existence. This is a double conditioning from Foucault's *Archaeology*, which addressed forms of thoughts and modes of speech rather than the ways of life, society or culture that were addressed more in Foucault's detailed historical investigations. It is also conditioned by the colonial archive, which recorded petitions, statements and queries, and assessments of the material environment, more than it did the actions and everyday lives of its people. Yet, this was a sphere in which people crafted their non-verbal enunciative modality, and traces of it are occasionally left in the archive. Simply through refusing to live in a certain way people could craft their identities and resist the hierarchies embodied in the city.

For instance, many people attempted to retain their lien on class D clerks' quarters in the early 1920s because they were cheaper than C class and still reasonably comfortable.[94] This refuted the principle that emoluments should be represented in housing, and that people would want to move up the hierarchy. Similarly, Rouse's report of 1929, when housing allowance was still in existence, showed that people would make applications they knew would be denied simply to get the extra cash.[95] It was also argued that people would apply for unorthodox clerks' quarters simply because there were more of them available.[96]

People could also resist the hierarchies of imperial space by crafting very personal places within them. A report from 1937 showed that during religious festivals or weddings, electric installations in government quarters had been hacked into to provide temporary street lighting.[97] This was claimed to be against the provisions of the PWD code, as well as hazardous for the health and safety of the occupants. A more common, and permanent, problem was the keeping of cows and buffaloes in government quarters. Lieutenant-Governor Sir Henry Gidney raised this issue in the Legislative Assembly in 1932, claiming that the cows were unsanitary, led to a profusion of flies and threatened New Delhi with an epidemic.[98] This was claimed to happen across all ranks of government servants. The issue was discussed again in 1937, when it was admitted that the milk supply in the capital was too low, but that clerks' quarters below grade B were too small for the keeping of cattle.[99] It had been suggested that to intervene would be classed as religious interference, which was brushed aside as irrelevant. A petition with 24 sheets of signatures was even submitted to the Viceroy, defending the right of people to bring their traditional lifestyle into the heart of the capital. The practice, however, clashed with Western notions of hygiene so it was continually campaigned against. By the late 1930s, however, this was the least of the Government's worries as it desperately tried to salvage an image of the ordered capital that had been designed two decades before. This period also saw re-orderings of the policing landscape, to instil discipline, and the urban fabric, to guarantee health, as the following chapters will show. The result, in New Delhi, was a series of schemes that sought to save the original image of the city, rather than substantially revise it to meet the changed material circumstances that it faced.

Re-visioning the conceptual landscape

As was shown earlier in this chapter, in 1929 the Government had admitted that the under-building policy had been a mistake, but proposed no immediate action with regard to building. Rather, this period saw the beginning of the reclassification of clerks' quarters to protect the officer class. It was also decided between May and July 1929 that clerks and newly arrived railway staff would be accommodated in the Western Extension.[100] The extension had been part of the capital project and had been intended to relieve congestion pressure on the old city (the consequence of this invasion is made clear in the third section of Chapter 4).

However, this period also saw a more comprehensive re-visioning of how the capital would be administered. This was not a thoroughgoing reshaping of the capital that affected the discursive objects it housed. Rather, it was rethinking of the concepts that, *mutatis mutandis*, would organise the succession and co-existence of objects in the city. Foucault (1972: 59) referred to such occurrences as 'procedures of intervention'; periods of rewriting or transcribing, or the translation of statements into new forms.

Future development and administration

Despite the mounting material, anecdotal, and archival evidence that the city needed 'progressive building', the reports that mark this intervention are saturated with protectionist and conservative rhetoric. Between 1928 and 1930 a 305-page file was compiled regarding the 'Future Administration of New Delhi'.[101] The scheme tackled the transfer of responsibility for the city from the central government to the NDMC. Mr J.N.G. Johnson, the Deputy Commissioner of Delhi, drew up the opening proposals in December 1928. Even though the NDMC members were carefully vetted, Johnson made it clear that the Government would not '... endanger the welfare of their new capital by making over control of important interests to a body which they cannot rely on or upon which, in last resort, they cannot exercise an ample check'. Despite arguments in favour of elective local self-government, Johnson insisted that for a winter capital, the Government should reserve the requisite control. Hinting at memories of the co-existent conceptual landscape of the cantonment, Johnson insisted that New Delhi should not be regarded as a 'normal town', but as an 'official estate', in which the concerns of the local taxpayer were second to those of the Government due to its financial investment in the city.

The extent to which powers could be handed to the NDMC was assessed by function; from buildings to roads, sewage, water, public health, education and lands. Johnson claimed to have saved this most important issue until last. It was Johnson who, as Chief Commissioner in 1936, would complain to the Government that Old Delhi was 'in the dark' regarding the activities in the capital region. Yet, this report was written at the time when the capital city's interventions into Old Delhi were being felt most intensely, whether through the extension of New Delhi's categories into the old city or through congestion in the walled city, caused in part by displaced workers from the capital. Johnson obviously saw this as an opportunity to rectify the balance of power between the new and old cities, the latter of which he stressed was in a 'mist of uncertainty' regarding the capital's development. While it was not suggested that the (Old) DMC govern the capital, the Chief Commissioner was suggested as the ultimate authority for the city, as Land and Development Officer. An interdepartmental conference was held on 29 November 1929 to consider Johnson's proposals. All of them were granted or agreed for future reconsideration apart from the Land and Development Officer. It was agreed that the post should be taken by the President of the NDMC, who would be a civil servant working in collaboration with the Chief Commissioner, but directly answerable to the Government. This was a vital opportunity missed by the Delhi Administration, for the landscape of the NDMC would now be administered not by those primarily concerned with the congestion of the old city, but with the finances, security and aesthetics of the new.

This was demonstrated in the proceedings of a meeting held to discuss the future of New Delhi in February 1931.[102] Following the inauguration ceremonies, Swinton, Lutyens and Brodie from the original DTPC attended the discussion and the minutes suggest a pre-occupation with maintaining the original image of the city. Byelaws were called for to regulate the design of private buildings, building lines, style and materials. While it was admitted that town extensions were needed, the emphasis was to lie on the prevention of disfiguring or unsightly buildings. Even this modest aim was not met. A meeting in March 1932 to discuss the progress made on the objectives set at the meeting agreed that any future growth of demand for clerical accommodation would be met in the City Extension Area.[103] This area was included to the south of the old city in the 1914 plans but never materialised. The land was used to house the Government of India Press, with the clerk's quarters that were provided being used to house the workers of the Press itself.

As such, the Government launched no major initiative to tackle the housing crisis throughout the 1930s. It was only in July 1938 that it was felt that New Delhi had developed sufficiently to require a stock-taking measure, in order to compare the material city to the planned form.[104] However, the emphasis still lay on the 'preservation and improvement of the aesthetic features of the new city'. Acknowledging that space for expansion was limited, the New Delhi Development Committee (NDDC) was charged with controlling, in a negative sense, not planning, in a positive sense, future development.

Two maps were provided with the NDDC report, showing the planned layout from the second report of the DTPC from 1913 and the actual development. Against the clear functional zones of the plan (see Figure 2.1), the actual city displayed a kaleidoscopic confusion of zones and functions. The Committee claimed that in 1923 it had been decided that the intention was not to create a new city, but a new quarter of Greater Delhi, devoted to the purpose of Government. Again, memories of the cantonment as a conceptual organisation of space emerged, although this time New Delhi was envisaged not as Military or Civil Lines, but as an administrational grid or bureaucratic cage. The Government's ownership of all the land meant any errors in development were due 'not to lack of power but lack of coordination'. A lack of communication with the Chief Commissioner was acknowledged although no structure was put in place to remedy this in the future.

The focus on the material landscape, rather than its relationship to human need, allowed the NDDC to claim that the administration of the land had been successful. There was barely any mention of the additional housing required, the information regarding which was contained in a separate appendix. However, this showed that the capital still required an extra 3,811 dwellings; 267 for gazetted officers and those earning more than Rs 600 per month, 1,863 for clerks and 1,681 for menial staff.

Middle-class housing colonies were considered and informal inquiries made with the Chief Commissioner, while private development was to be further promoted. For official residences, bungalows were planned in patches throughout New Delhi, taking in peripheral land near the race course, Willingdon Crescent, the civil aerodrome and Lodi Road. The orthodox and unorthodox clerks' quarters were also planned for similar areas, marking a radical blurring of the previous spatial hierarchy (see Figure 2.2).

Ribbons and rents

On 16 July 1940, the Viceroy's Private Secretary had written to Chief Commissioner Askwith stating that Viceroy Linlithgow was taking a keen interest not only in the outcome of the NDDC proposals, but also in actions to target ribbon development along the roads around New Delhi.[105] Building on land outside the limits of the capital region was one of the responses by people who had been forced out of the local housing market by rising rents, which themselves had responded to the massive outstripping of supply by demand. While the Government had been slow to cure the housing shortage, it did devote its energies to treating the symptoms of this condition.

As far back as 1913, merchants living in Delhi had complained about the rapidly rising rents. The Chairman of the Punjab Chamber of Commerce had written to the Chief Commissioner on 22 February 1913 that, 'It is obvious that unless sufficient space is allotted for the building of bungalows which will be available for non-officials a fictitious value and high rents will at once follow on the fixing of the site for the new city.'[106] As such, the IDC had insisted in 1913 that it would under-build only on the assumption that the Government would build further accommodation should rents be forced up. However, it has been shown that the Government did not respond to this commitment, and it was pointed out in the Legislative Assembly in 1937 that the many unhoused government workers were paying exorbitant rates.[107] A year later the Viceroy expressed his first concerns about ribbon

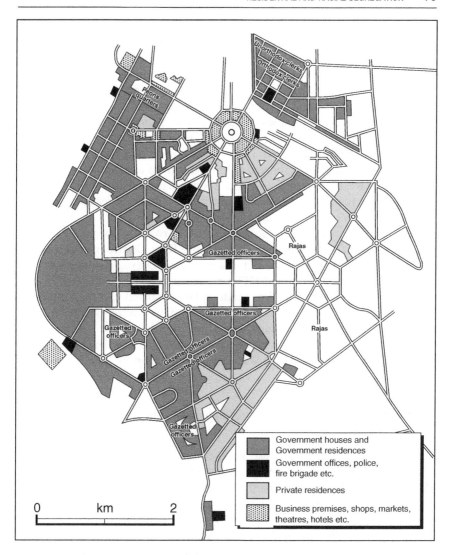

Figure 2.2   Actual layout of New Delhi and the use of land as in 1938

development to the south of New Delhi and measures to target the two symptoms of the housing crisis developed in tandem.

In response to the Viceroy's concern, Chief Commissioner Jenkins asked A.P. Hume, the Chairman of the Delhi Improvement Trust (DIT) to look into the matter.[108] Although the DIT (see Chapter 4) was focused mainly on the old city, in March 1938 Hume advised against extending the NDMC limits, but proposed a law forbidding construction on or near certain roads.

In a further move away from *laissez faire* political economy, a year later the New Delhi Rent Control Order (1939) was issued.[109] Powers were allotted to the Government to target any rent within New Delhi or the Civil Lines that was deemed 'excessive'; a 'fair' rent would be set and enforced for the building. Given that most of the capital's unhoused population lived in Old Delhi, the Order granted the image of control to the capital but not the benefits of control to those most in need. After the first year of operation, it was reported that rents were being calculated varying with area, capacity and ground rents.[110] By December 1940, 1,122 orders had been put in place, which were said to have stabilised rents in the areas affected.

The Delhi Restriction of Uses of Land Act was passed on 8 April 1941.[111] Land within 440 yards of a named road within Delhi Province could be controlled, in terms of buildings and means of access to the road. The Chief Commissioner immediately notified a series of roads to the south of New Delhi. In August 1942, the New Delhi House Rent Control Order was further extended to the Western Extension and to Daryaganj, in south Old Delhi, as both areas contained DIT properties and were occupied by government workers.[112] Complaints made in 1943 showed that the Order was being abused and used *against* tenants as landlords were evicting tenants and then re-letting at the original rate a few months later.[113] However, by this time the housing market had been elevated to a new level of crisis by the onset of the Second World War and the transition of New Delhi into a war capital.

C'est magnifique, mais ce n'est pas la guerre ('It is magnificent, but it is not war')

Britain and France declared war on Germany on 3 September 1939 and within a year the local economy of Delhi had been transformed. The Estates Officer of New Delhi wrote to the Chief Commissioner in October 1940 regarding the even higher rents that had emerged.[114] As opposed to the target rent of 10 per cent of salary, many were paying 20–30 per cent, while rents in certain districts such as Daryaganj had risen by 50 per cent. In addition, travel costs, doctor's fees and tuition fees had all increased rapidly.

Yet, war also left a stark impression on New Delhi's material landscape. The Government had to coordinate the war campaign across the whole of British India, which required a huge increase in the number of personnel and office space. The Government acquired the Princes' Palaces and built temporary offices in many of the empty spaces around Kingsway. An undersecretary living in the heart of New Delhi wrote that, 'It is remarkable to see the spread of the Govnt. Offices like a kind of tumour, into Princes houses. Kashmir, Hyderabad, Travancore House, + possible others, are taken over and the ants run ceaselessly to and fro in their marble halls. C'est magnifique que [*sic*] mais ce n'est pas la guerre, one is inclined to say. Otherwise Delhi knows nothing of war.'[115] Viceroy Linlithgow wrote to the King on 5 February 1940 regarding his unease about the direction that building was taking in the capital.[116] By 1943 over 1.2 million square feet of official accommodation had been provided, in addition to residential accommodation for 969 officers and 745 clerks.[117]

The collapse of the spatial hierarchy that this construction, along with the post NDDC plans, brought about was mirrored in a complete social shift. New Delhi society had drawn sustenance from, and been modelled on, a nostalgic view of England. This England had long since collapsed, if it had ever existed at all. The colonial environment had fortified socio-spatial boundaries, whether grounded upon cultural, epidemiological or eugenic fears of mixing and the collapse of order. But the influx of troops and out-siders into New Delhi shattered the social rigidity of the interwar years. Viola Bayley recalled that the Second World War ended 'old India'. When she had arrived in India in 1933 women could not leave the house without a special hat or appropriate dress, yet by the end of the war women could ride bikes without stockings or hats: 'Kipling's India of parasols and gloves and ladies lying on chaises-longues were finished. It was the troops who exploded the sun-stroke myth. Everywhere one saw them off duty, hatless and stripped to the waist in sun that a few years before would have been considered deadly.'[118] In these Indian summers of topless Tommy's and stockingless Viola Bayley's, the rigidity of the Warrant of Precedence was eased and the social tensions within the capital became obvious to see. As Aldous Huxley (1926: 138) had previously observed:

> The comedy of Delhi and the new India, however exquisitely diverting, is full of tragic implications. The dispute of races, the reciprocal hatred of colours, the subjection of one people to another – these things lie behind its snobberies, convention, and deceits, are implicit in every ludicrous antic of the comedians.

Yet, this was as much a product of much more deep-seated changes. The Indianisation of the Civil Service had maintained the rank, but disturbed the race, of colonial hierarchies. Anti-colonial nationalism had also demolished the supreme self-confidence and belief of the Raj in its beneficent and benevolent nature. In local terms, this was represented by the increasing weight of Old Delhi on the minds of the capital, as explored in Chapter 3.

Chapter Three

Disciplining Delhi

This chapter explores the genealogy of urban discipline in New and Old Delhi. It seeks to expose the ways in which attempts were made to order Indian bodies to make them politically docile and productive of a secure capital. The government was continually pushed to increase its expenditure on the military and the police due to the resistance of the local population, whether through crime, complaint or outright revolt. Across the two cities can be traced a gradual and halting shift in the regime of government (see Table 1.1). *Epistemologically*, the faith in calculation and planning increased, bolstering the belief that disciplinary mechanisms could make the city safe and knowable. The presumed *identities* of the local population shifted from general untrustworthiness to a fortified belief in seditious anti-colonial sentiment and communally defined religious fervour. This provoked a need to both *visualise* and know the city, which was serviced by a series of re-mappings. These sought to understand the geography of urban risk and to distribute surveillance and punitive forces throughout the landscape. These forces were equipped with technologies of discipline and were given ever more sophisticated training in using them, forming a *techne* that was more violent and intrusive than the European model on which many of the practices were based. Finally, the *ethos* established in 1911, that of protecting the capital, seeped into the practices not only in New Delhi but also throughout the older city.

This disciplinary regime of government will be analysed in three sections, which successively work their way from the heart of Empire in the new Delhi into the supposed heart of Darkness in the old. The first section will deal with the policing of New Delhi that sought to prevent crime in the core and the invasion of the periphery. The second section will look at the evolution of disciplinary mechanisms, embedded in the law, for reinstating order in Old Delhi in times of riot or anti-colonial uprising. Finally, the third section will consider the more sophisticated diagramming of urban order in the old city during religious festivals and riots.

New Delhi: Policing the Heart of Empire

Policing: From government to discipline

The emphasis, in this chapter, is on the police as agents of discipline, although the police's European origins hint at a more governmental remit of operation. 'To police' (Foucault, 1978a [2001]: 206–7) society originally meant to administer the nation like a home, knowing its details and content, in line with the cameral science of the police in seventeenth- and eighteenth-century continental Europe (Hindess, 1996: 119). Here, the police were responsible not only for public order but also for morality, common measurements, trade and landscaping, as a replacement for the regulations of the feudal period (Pasquino, 1991: 111). Later, liberal arts of government also relied upon the police to enforce public codes of civility, reason and order, and private levels of self-mastery and command. The nineteenth century saw the emergence of police forces whose intense local knowledge of towns and citiesallowed them to operate '. . .not so much through terror and the certainty of apprehension, but by placing a grid of norms of conduct over urban space and regulating behaviour according to the division of the normal and the pathological' (Rose, 1999: 73).In addition to regulating social norms, the police were also agents of liberalism in the other domains of government. In terms of the economy, fraud and tax evasion were investigated. With regard to biopolitics, standards of habitation and reproduction were enforced; at its most extreme, linking the police state and biopolitics through eugenicist and racialist programmes (Barret-Kriegel, 1992: 193).

However, such contextualisations within a broader governmentality must admit the indissociable presence of disciplinary power formations within the police. Colin Gordon (2001) emphasised the continuities between the confessional urge of pastoralism and the surveillance of disciplinary institutions. Those people who failed to regulate their conduct in any of the liberal realms of government could be submitted to disciplinary practices to normalise their behaviour. Police, as such, function as the technological apparatus of biopower rather than servicing either its micro- or macro-poles. Although the police in Delhi were shown to seek 'moral effects', this chapter will focus on disciplinary acts in urban space as practised at times without the liberal constraints of Europe. The specifically colonial and urban nature of the policing in question will be examined after a brief exploration of Foucault's writings on discipline.

Overtaking the sovereign model of power as the primary power relation amongst certain bodies of the population, the arts of discipline sought
to manage individual bodies, rather than being concerned with territory
and making periodical demands for tax or men. Discipline would later be
advanced most wholeheartedly by the bourgeois classes (Foucault, 1975–6
[2003]: 36) and came to be embodied in the great institutions of the clinic
(Foucault, 1973; Philo, 2000), the workhouse (Driver, 1993), the asylum
(Foucault, 1967; Philo, 1989), the jail (Foucault, 1977; Ogborn, 1995) and
the lock hospital (Foucault, 1979; Howell, 2000). By creating the impression of constant surveillance, disciplinary tactics sought to interpellate a
self-disciplining subject that would be politically docile yet economically
productive and socially secure.

While Foucault detailed four arts of discipline concerning space, activity,
exercises and combinations of forces, he announced that 'discipline is above
all an analysis of space' (see Driver, 1985; Elden, 2001: 139). Disciplinary
space was described using four arts of distribution. The first was that of
enclosure, whether it be the monastic model or that of the school, barrack
or factory. Yet discipline only 'sometimes requires enclosure' whereas the
principle itself is 'neither constant, nor indispensable, nor sufficient in a
disciplinary machine' (Foucault, 1977: 141). More important is the art of
partitioning, in which each individual is allotted a place in order to judge,
know and use them; the creation of analytical space. This space is made
both administrative and political through allotting sites a function, whether
it be surveillance or production, such that the performance of the individuals
posted there can be assessed. The result of this function allows the fourth
art, ranking, which compares one space to the next or to a norm. These
four arts of distribution, together with the dictates on activities, time and
composition, supposedly created the impression of constant surveillance
without the need for constant supervision.

The organisation of these arts of discipline that has been studied most
intensely has been that of the Panopticon, 'the diagram of *a* mechanism of
power reduced to its ideal form' (Foucault, 1977: 205; emphasis added).
The fact that this is *a*, not *the*, mechanism of power indicates that there
has been, and will be, different diagrammatic representations of disciplinary
distributions of subjects and objects (Hannah, 1997: 172).

Foucault noted three processes that were essential in facilitating the movement of the panoptic diagram of power from an inward facing blockade,
the Panopticon, to an outward facing mechanism, that of panopticism. In
addition to the 'functional inversion of the disciplines' and the 'swarming of disciplinary mechanisms' came the 'state-control of the mechanisms of discipline' (Foucault, 1977: 213), namely, the institutionalisation
of the disciplinary techniques of surveillance within the administrative
machine of the police. The direct gaze of the Panopticon was replaced

by surveillance: both directly through urban patrols and indirectly through files and dossiers, creating 'an interstitial discipline and a meta-discipline' (Foucault, 1977: 215).

The history of the British police is dominated by the formation of the London Metropolitan Police Force in 1829, and the reform of the boroughs in the mid-1830s, as a reaction to urban 'mobs'. This was not only a move to secure economic forces, but also public space, state structures and social norms (Gatrell, 1990). In addition, both Ogborn (1993b) and Fyfe (1991) have stressed that the locality of implementation was just as important as the centre of innovation. However, fewer studies have looked at the colonies, rather than the counties, as 'periphery'.

Brogden (1987) has, however, examined police formation with regard to the colonies, where the Royal Irish Constabulary model was used due to its control by civil authorities and links with the military. Arnold (1986) had corroborated this view in his analysis of Indian policing in colonial Madras. The diverse services and armed, centralised forces made the Irish model more suitable for a system that supervised both the subject population and the lower ranks of the force itself. A common feature of both the Irish and Indian police was the more regular use of violence and force, as against the disciplinary use of surveillance and coercion. This was framed as a necessary evil given the lack of self-discipline of the colonial populations (see Anderson and Killingray, 1991, for comments on the specificity of each colonial police force and a critical commentary on the Irish model thesis).

Just as the whole of nineteenth-century society could not be enclosed within a disciplinary institution, nor could colonial urban India be placed under the constant state of the 'counter-city' (Foucault, 1977: 205). Instead, discipline was enforced by the interstitial institution of the police. At this level of investigation not only do the mechanisms of disciplining a non-institutional population become clear, but so do the means for resisting or avoiding the disciplinary gaze. Indeed, it was often the internal reviews that followed a period of problematisation that defined how disciplinary technologies were deployed, although each new deployment necessarily created as many spaces of resistance as those it made visible.

Policing the capital

This recrudescence of outrages has caused widespread alarm among the ministerial staff of the Government of India and I am to request that effective measures may be taken to protect the residents of the New Capital from outrages of this nature. (Imperial Secretariat Association to the Chief Commissioner of Delhi, 11 January 1927)[1]

Within two months of the Secretariat being occupied, New Delhi found itself in the midst of a swelling crime wave. The ISA voiced the concerns of its members (see quote above). It cited the theft of a safe from a local market shop on 29 December 1926, the breaking open of safes and *almirahs* (chests) in the post office in the North Block of the Secretariat in January 1927, attempted burglaries from five different squares of clerks' quarters and a theft from 10, Queen Victoria Road in the very heart of the city. The Rail Board also submitted a petition signed by 30 people pleading for protection, claiming that *dacoits* (armed robbers) were regularly entering workers' houses.[2] Worse was to come in April 1928 when the home of the Accountant General, Central Revenues, was invaded by *dacoits* who ransacked the entire house in the presence of the inhabitants.[3] Even the Army voiced its concern, reporting burglars armed with swords and *lathis* (staffs) in the heart of Raisina and a population that 'passed nights in fear'.[4]

These crimes targeted not only the spatially peripheral, numerous, but poor clerks, but also the central, dispersed, but wealthy officers. It was a targeting not determined by race, for both Indians and Europeans were victimised, nor was it dictated by class, for both the poor and the rich were victimised. It was crime drawn to a landscape of opulent wealth and a spacious layout that facilitated rapid escapes. It was also crime that occurred in the sparsely populated city before its formal inauguration, but which continued throughout the province for the next 15 years. Annual non-violent offences increased from an average of 1,572 in the 1920s to 1,910 in the 1930s, rising exponentially during the war years to 5,604 cases in 1943.[5]

Yet, violent crimes were also on the increase. On 8 March 1931, dacoits armed with guns and knives attacked a government employee in the clerks' quarters of New Delhi, following a similar attack on 7 February, during the actual inauguration of the capital. A question was posed by R.K. Shanmikham Chetty on 21 March 1931 in the Legislative Assembly asking what the Government intended to do 'to prevent such occurrences from time to time in the Government of India Headquarters?'[6] Following a series of violent burglaries, another question by Mr M. Maswood Ahmad on 21 October 1932 asked whether the Government was aware that certain quarters further removed from New Delhi were 'frequented by thieves' but not by the police?[7] The reply was that there simply was not enough police staff.

There was growing anxiety in both the local population and the Government about crime rates in the capital. In a Legislative Assembly debate, the government was forced to admit that between 1 January 1939 and October 1940, the 1,284 burglaries and 2,331 thefts in Delhi amounted to Rs 432,567 of property lost, of which only Rs 31,404 (7.2 per cent) was recovered.[8] Although the total number of serious offences in New Delhi stations amounted to 577 in 1941 in comparison to Old Delhi's accumulative total of 1,320,[9] mapped onto the geography of population density this represented a crime for every 162 people in New Delhi, but only for every 395 people in Old Delhi. Although Delhi Province obviously had a geography of crime, it also had an internally administered geography of policing.

Throughout Delhi, policing staff were lodged at either police stations or posts. An analysis of the distribution of police employees in Delhi Province illustrates the extent to which New Delhi was in fact *over*-staffed in terms of policemen per head, compared to Old Delhi. However, this police force was used to protect certain parts of New Delhi while attempting to survey and regulate the older city.

In looking at the spatial distribution of police staff, there is initially no huge disparity. The area bound by the Delhi Municipal Committee (DMC) was policed by 501 policemen, 47.9 per cent of the provincial police force, while the New Delhi Municipal Committee (NDMC) area employed only 223 policemen, just 26.1 per cent of the workforce. Indeed, in terms of policemen per square mile, the DMC was much more intensely policed, having 23 policemen per square mile compared with 14.2 in New Delhi (only the Cantonment had a lower rate, explained by the high army presence). However, this geography of distribution is wildly distorted by the geography of density. The DMC's 21.74 square miles contained 521,849 people in 1941, compared with the 93,733 in New Delhi's urban 48.3 square miles, or the 19,395 in the highly dispersed 8.86 square miles of the Civil Lines, the alternative site for Europeans and home of the Delhi Administration.[10] This gave the DMC a population density of 24,004 per square mile, but with only 23 policemen for the same area, while New Delhi had a population density of 1,939, with 4.6 police staff for each square mile. As such, in Old Delhi each policeman had to monitor 1,041 people, whereas in New Delhi the ratio was 1 policeman to just 420 occupants (see Table 3.1).[11]

While the policing of the old city will be examined in depth in the next section, there were significant variations *within* New Delhi that also illustrate the priorities of the policing administration. The tension between protecting the few, high profile houses south of Kingsway and the many, less

Table 3.1   Relation of police, population and area in Delhi

	Percentage police force	Percentage population	Percentage area
DMC	47.9	75	3.8
Civil Station	12	2.8	1.6
NDMC	26.1	13.5	8.4
New Cantonment	2	3.3	2.9
Delhi Province	12	5.4	83.3

prestigious residences north of the ceremonial path becomes apparent. The Tughlak Road Police Station that protected the 20,000 residents south of Kingsway employed 51 staff, whereas the New Delhi Police Station and Reading Road Police Post employed 147 staff for the 80,000 in their jurisdiction in the north. Every police worker in the elite area south of Kingsway effectively had 392 people to police, whereas in the clerks' quarters in north New Delhi, each worker had 544 people to safeguard. As such, of the 158 crimes reported from within the clerks' quarters of the north-west part of New Delhi in 1941, only 27 of them (17 per cent) ended in conviction.[12] The government's suggestion that Delhi simply did not have enough staff to fulfil its duties should instead be read that those staff had certain duties to fulfil, and areas to protect, before others.

This was pointed out in the Legislative Assembly during 1940–1, just as the crime wave sweeping the capital was becoming apparent. M.S. Amey asked whether the Government was aware of the great insecurity prevailing among clerks and assistants due to the frequency of burglaries and *dacoities* (armed robberies) and the murder of a servant during a burglary on 24 September 1940?[13] A further question asserted that the clerks were afraid to leave their houses and requested that police be removed from the Legislative Assembly itself to tour quarters of the clerks and peons. Later in the year the Home Member had to admit that most of the thefts and burglaries in the capital were committed against subordinate staff when the men were at work.[14] The clerks' quarters, it was admitted, were patrolled by just 1 head constable and 8 foot constables by day, and 24 by night. This dimension was raised again in October 1941 when the Government was asked which localities were most vulnerable in terms of burglaries and theft, and what was being done about it.[15] The targeted area was admitted to be the clerks' quarters around Reading Road, and that more intensive patrolling was being considered.

Although the total crimes reported to New Delhi Police Station in 1942 amounted to only 26 per cent of the Provincial total (1,036 of 3,928 crimes), it was this part of the city that was most intensively reorganised.[16] The reorganisation marked an attempt to discipline space through partitioning the landscape, allotting police patrols to increase surveillance and communicate this information back to the police station that could assess the need for further action. This reorganisation followed a letter being sent from the Senior Superintendent of Police to the Chief Commissioner on 20 May 1943.[17] This prompted not only the upgrading of Reading Road Police Post to a Police Station, but also the intensification of the patrol rota for the New Delhi Police Station section of the city. The most elite areas of the city to the south of Kingsway had already been reordered, but this was in response to the threat of extremist attack rather than crime, as detailed in the following section.

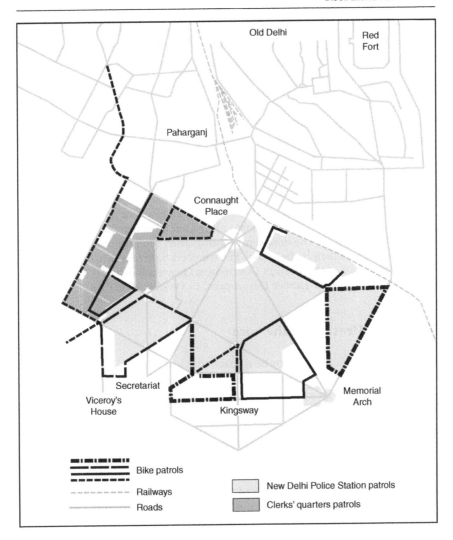

Figure 3.1   Police patrols in New Delhi, 1943

The city was divided into patrol areas, each to be guarded by three foot constables and cycle patrols, as illustrated in Figure 3.1. It would be unfair to say that the rest of New Delhi was completely neglected in favour of the central bungalow zone. However, the Reading Road (clerk's quarters) area was organised into 12 day and night beats, 2 single cycle patrol routes and 1 double cycle patrol route, whereas central New Delhi was formed into 27 beats and 6 cycle patrols. Although the area was larger for the central

New Delhi beats, the population was much less dense. As a centre of both economic and symbolic capital, the commercial complex of Connaught Place received massive attention, being protected by eight beats of three foot constables. The majority of the police force would be occupied with protecting the parts of New Delhi reserved for the officers, not clerks, of government.

Despite this, complaints and crime continued throughout the 1940s. Using terms reminiscent of those complaints 17 years previously, it was clear that for the lesser privileged members of the government, very little had changed:

> There have been many thefts in the locality and not a week passes without some robbery or theft taking place. Complaints to the police have not resulted in any improvement of the situation ... If the family of the clerks just go out for a few hours to the city or to some other quarter, you may rest assured that there would be a theft in the house. The members have no peace of mind and are greatly afraid.   (Karol Bagh Government Quarters Residents' Association to the Secretary of the Labour Department, 17 October 1944)[18]

The Keep: Protecting the core

New Delhi has been shown to be a space that was intensely policed, yet this attempt to control the urban environment was by no means successful. This policing displayed elements of partition, through patrols, of functionalisation, through the surveillance for suspect characters, and ranking, through the police stations (compare with the surveillance and ranking of social class discussed in Chapter 2). There was no obvious sign of enclosure, the first art of disciplinary distribution. Yet, New Delhi itself existed as a form of enclosure, separate and protected from Old Delhi to the north. As such, the capital had not only to be protected from a stealthy enemy within, but also from the possibility of attack from without. The response was two-fold; to protect the core and to defend the boundary.

As has been shown, the Government of India and the European population at its heart formed the core of New Delhi. Between 1930 and 1932 when Delhi was faced with rising crime, the Civil Disobedience movement, and the inauguration of the capital, the sufficiency of the police to protect the population of New Delhi was called into question. An internal memorandum of 1929 from the Home Department to the Chief Commissioner of Delhi stressed the threat not just of crime, but also of revolutionary attacks, in the capital region.[19] Following the attempted assassination of Viceroy Hardinge during his state entry into Delhi in 1912, the province had suffered various extremist attacks. In 1928 the Hindustani Republican Socialist Army was reformed and organised the high-profile bombing of the Legislative

Assembly in April 1929 (Lal, 1999: 148). The Government urged greater protection for the Viceroy and Home Member, although the initial reaction was to appoint a Bengali Inspector in New Delhi on the presumption that the extremists had come from the more radical climes of Bengal.

Six months later a bomb was detonated under the Viceroy's train as it passed the terminal point of Kingsway in New Delhi, although the train was not derailed or damaged. The need for more close-knit security was realised so pickets and patrols were established in the most elite areas around and to the south of Kingsway. These predated the patrols that would only be introduced at such an intensified level north of Kingway 13 years later. There was also an additional motor patrol that took in a wider sphere of surveillance touring the route in an alternate clockwise, anticlockwise pattern.

The Civil Disobedience movement of 1930, though centred on the old city to the north, unnerved the civil servants to the south. This was combined with continued extremist plotting, leading the Deputy Commissioner to claim on 31 December 1930 'that for some months past Delhi has been a centre of terrorist activities and that the danger of an outrage is serious. A spectacular crime in Delhi would give a stimulus to the terrorist movement as a whole and would damage the prestige of Government'[20]. A review of police procedure for protecting New Delhi was undertaken in the late 1930s in which the priority was always made clear to be that of protecting the central spaces and residences. The Secretariat themselves were organised in December 1930 to make points of entry more visible through increased patrols, lighting ('worth several policemen') and pickets. The Chief Commissioner agreed with these proposals if the Government was willing to pay for them themselves.[21]

Throughout the province, the protection offered was very much directed by the concept of prestige, with 18 extra head constables and 96 foot constables being allotted to protect key sites in New Delhi, yet only 6 head constables and 29 foot constables for Old Delhi. These were in addition to the extra 1 Inspector, 2 Sergeants, 50 head constables and 100 foot constables who had been sanctioned for the Government of India's protection from October 1930. Calls for more police would continue to come from the Government, until the Chief Commissioner had to demand, on 18 November 1931, that no more police would be requested unless the Government was willing to pay for them.[22] Despite this, by 1943 the Viceroy's House Guard alone totalled 126 men.[23] Although these measures were thought ample to protect the population and government in peace time, the need to protect the capital in the event of a wider outbreak of violence also had to be considered.

In the event of an all-out invasion of the capital, there were plans to withdraw the trustworthy civilian members to a safe haven that could be defended. The designated safe place ('The Keep') that was easily defended in Delhi was the Fort in the old city, later to be supplemented by the Secretariat in New Delhi, to which the vulnerable population would be evacuated in times of need. The perimeter of the Keep was divided into Defence Sectors, concentrating the surveillance urge from the whole of New Delhi down onto the Keep itself, as it would later be extended over Old Delhi.[24] A garrison was allotted to each sector in proportion to the buildings to be defended, such that it could cover any attack with both enfilade (sweeping) and oblique (angled) gunfire. The defence was organised from the 'final Keep' in the wired and loop-holed central block of buildings. The civilians were reclassified into a contained workforce, and duties were divided between those who could and could not bear arms, and then by sex. Women contributed to cooking, medical work and administration while men focused on engineering and accommodation work. The rigorous planning of the Keep's defences, arms, administration, supplies, communications and alarms came to nought, as the Keep was never used in Delhi. Yet, the debates about its nature and contents provide a window into the *identity* assumptions underwriting the practical rationalities of government security.

In planning the Keep, concepts of prestige and safety had to be held in balance. The discussion regarding how bodies were to be ordered in space was implicitly grounded on the assumption that without the ordering and inherent rationality of the white man, Delhi would descend into atavistic clan warfare and pandemonium. In 1930 it was suggested to Brigadier E. Giles, who had to plan the military defence of Delhi, that civilians should be evacuated to the Keep under what he considered a relatively low level of risk. He responded, on 6 June 1930, that:

> [t]his would presumably mean that all officers and civil officials residing in New Delhi would be required to concentrate in the Secretariat. The news that the Government of the Indian Empire had been driven by a half-armed mob to take refuge in their offices would spread like wild-fire from one end of the country to the other, and all its prestige would vanish in twenty-four hours. Looting would start at once, communal feuds would blaze up everywhere and villagers would seize the opportunity of paying off old scores against their neighbours.[25]

However, another set of advice to the Government suggested that New Delhi was not a self-contained unit and the ordinary police would not be able to defend it. Indeed, while Eric Hobsbawm (1973 [1994]: 224) claimed that few cities could be less suitable for a riot than New Delhi, the spacious layout actually made an influx of peoples into the city hard to detect. Military arrangements that allowed people to stay in their homes were decided to be desirable though not essential, but New Delhi did need it's own Keep. By 1938 the 'New Delhi Secretariat Keep' was confirmed as the safe haven for government workers plus their families who lived in New Delhi, although the families of workers who lived in Old Delhi were to be given no protection.[26]

In addition to the fear of revolutionaries and extremists, there was also a seeping ambivalence within the Government as the influence of Gandhi and the Indian National Congress spread. The fear was not simply one of the debased, wily Oriental, as Said's (1978) divisions often suggest, but of mimicry and fracture. People may have worked for the Government, but this was often just for money or opportunity; even the most loyal would remain, as Bhabha (1994: 86) claims, 'as a subject of difference that is almost the same, but not quite'. This was a long acknowledged anxiety at the very heart of Government. A secret letter from the Home Department to the Viceroy Lord Hardinge, on 26 June 1912, bemoaned the lack of respect in Indians who had grown up under modern rule. Replacing the deference to the white man was an insolence that was 'significant of the true inner feelings of Indians who have some education if the restraints of official fear or favour are not operating, or if the relations of host and guest, or of personal friendship, are in question'.[27]

Compounded under the pressure of a growing nationalist movement, this distrust of Indian workers defined them more generally as a security threat. A memorandum was circulated on 14 May 1930 reminding Government workers that taking part in or aiding subversive political movements was against the Public Servants Conduct Rules.[28] Despite warnings, participation was again noted in 1932, but the impossibility of constant surveillance remained. Even if an officer from each department were sent to meetings to spy out nationalist campaigners, he would need someone to corroborate his statement: meaning two people would be required to track every one possible government traitor.[29] The Chief Commissioner admitted this was impossible and simply asked that some people be made examples of.

In determining who would be admitted to the Keep in times of crisis, this tension re-emerged. In 1940 the Government Departments and the Delhi Administration were asked to downscale their estimated Keep populations. These 'refugees' were defined as:

(1) all Europeans and their families;
(2) gazetted officers and their families;
(3) Anglo-Indians whose lives are in danger due to the nature of their occupation or methods of living;
(4) Indian subordinates who by the nature of their duties have taken an active part in the suppression of disturbances and whose lives would, therefore, be in danger;
(5) families of (3) and (4) living with them.[30]

The scheme was inclusive on race, all whites were protected, and clearly differentiated a superior class based on rank, the gazetted officers. Cultural and racial affiliation to the British was acknowledged in accepting hybrid Anglo-Indians, although the definition was left open with regard to an assessment of the danger of their situation. However, non-white (race) and subordinate (class) Indians were not accepted automatically, as they were in 1938.[31] Instead, only those whose work against the nationalists would place them at risk outside were accepted, with this figure being capped at 50 per cent of the workforce. However, despite these measures decreasing the size of the Keep from 20,842 to 16,349, 93.6 per cent of the approved refugees were Indian. Through stipulating that the only lower class Indians to be accepted would be those in danger *outside*, it was hoped that they would not be a threat *inside*. As a microcosm of the city and society, the Keep embodied the fears and ambivalence at the core of colonial culture. The Keep itself was never brought into use. The means to protect the city's boundaries, however, were.

Picket and patrol: Protecting the glacis

[T]he European and essential Government communities are so large that the only effective course is that which has been adopted; viz. to make New Delhi and the essential portion of the Civil Lines safe by measures to cordon off the city and by providing intensive police mobile patrols throughout the areas to deal with the isolated malefactors who may infiltrate past the cordon. (Chief Commissioner of Delhi to the Home Secretary, 18 September 1942)[32]

To the east,New Delhi faced the Jumna River, to the south the barren Delhian plains, interspersed with ruins and villages, to the west rose the Ridge that shielded the army Cantonment. To the north-east, beyond the Government of India Press and the *cordon sanitaire* lay the walled city. To the north-west lay the suburb of Paharganj and the extensions of the old city; Sadar Bazar, Sabzi Mandi and Karol Bagh. It was from the north that any attack against the capital would be made. There was also the possibility that the rioters would head north to the Civil Lines, attacking those who remained in the old European station. The bulk of preparations were for quelling the disturbances in the Old City itself, but there always remained a focus on defending the boundaries of those areas occupied by Europeans.

The interaction with the old city will be covered in the second section, but it is important to stress that the priority was always to protect the populations north and south of Old Delhi. In 1928 a revised version of the Internal Security Scheme, which detailed the military security plans, was rejected for offering insufficient protection for Delhi as the seat of the Government of India.[33] The area on 'prestige alone' was claimed to be of supreme importance and the police were described as inefficient to maintain the functioning of government in the case of an attack. As such, the Colonel Commanding the Delhi Independent Brigade insisted on 23 January 1928 that 'the problem of the security of the Central Government at Delhi should be reconsidered *ab initio* [from the beginning].'[34] Army Headquarters resisted any tailoring of military response to specific features of Delhi's urban landscape, claiming that the general principles of internal defence would suffice, but the Delhi Brigade continued to insist on the need for more troops.

In 1930 it was rumoured that a march was going to be made by Congress supporters to launch a non-violent protest outside the Viceroy's House in New Delhi. This prompted an internal discussion on the government's planned reaction. As the Home Department made clear on 2 April 1930: 'As regards the suggested move on the Viceroy's house, it is a cardinal principle of the Delhi administration that no crowds or procession shall be allowed to visit New Delhi or the Notified Area (Old Civil Station). Should a body of people start with evil intentions for the Viceroy's House they would be stopped at some suitable point between New and Old Delhi.'[35]

This principle was carried forward to the Quit India uprising in 1942, the next and final major mass movement before Independence. The Senior Superintendent of police's preparations reiterated that the priority was to protect the elite areas to the north and south of the old city and to contain any small packs of protestors that penetrated the barrier. On 9 August, the Deputy Commissioner made provisions for Criminal Procedure Code 144, banning public processions of more than 10, but only 'in case a determined effort is made to enter New Delhi. This order is *not* to be promulgated unless the crowd shows a determined effort to press on to New Delhi'.[36]

News of the arrests of senior Congress members began to leak into Delhi on 10 August 1942. The Senior Superintendent of Police later recalled that when 'it was clear that demonstrations might be made to enter New Delhi, fresh civil dispositions were ordered based generally on the Anti-Government Agitation Scheme'.[37] In terms of protecting the new city, this involved amassing 105 foot constables, 9 head constables and 4 Superintendents around the walled city to prevent the intrusion of any troublemakers and amassing 90 foot constables, 9 head constables and 9 Superintendents around Connaught Place.

After the outbreak of serious violence in Chandni Chowk on 11 August the Military Commander of the Delhi Area took over the operation using a 'greatly strengthened' form of the Annexure Local Alarm Scheme 'providing for the complete cutting off of Delhi City from New Delhi and the Civil Lines Notified Area by military pickets and patrols, the patrolling of Chandni Chowk and other important areas in the city.'[38]

On 18 August, the police resumed control of the city but kept the pickets in place. None of these pickets were inside the city; rather, they served to protect the enclaves that contained Europeans. Indeed, it was stressed that '[t]he duties of these pickets will be to hold up by fire power if necessary any crowd which may seek to leave the city by these routes to enter New Delhi or Civil Lines, giving immediate information to the nearest police station'.[39] Although more dispersed in the north, between 60 and 65 foot constables were allocated to protect both New Delhi and the Civil Lines. However, the main exit point from Old to New Delhi, Ajmeri Gate, was given special significance, the only site to have an extra head constable and a Senior Inspector. Also, as apposed to the army troops employed there before the police, the 153 Parachute Battalion had guarded this site.

Likewise, the police patrols by lorry and car that had been established on 12 August patrolled the walled boundary of the city, especially the two main exits, but also circled Paharganj, protecting the north-west flank of the city and the entrance to New Delhi by Chelmsford Road. Indeed, it was explicitly stated that '[t]his patrol will particularly look out for any attempt by demonstrators to infiltrate into New Delhi by this road' (Senior Superintendent of Police, 12 August 1942).[40] These patrols, unlike the static and vulnerable pickets, were mobile and penetrated the old city. They tied the management of the old Delhi to the defence of the new, reminding us that the bounded and segregated cities in Delhi were in fact part of the same capital region, bound together not just by name, but by administration and control. Yet, there was something profoundly symbolic and fearful about the boundary between the two cities, a boundary that in times of crisis distinguished self from other, those to fight with and those to fight against (see Chapter 4 for a discussion of this space as both a disciplinary and a biopolitical buffer zone). Yet, this was a fight not just at the gates, but one that was taken deep into enemy territory, as the second section will illustrate.

Anti-colonial Nationalism and Urban Order

The arts of discipline had to be more comprehensively applied to Old Delhi, in line with a larger population, a denser urban landscape and a collection of organisations tirelessly working to stoke anti-colonial sentiment. As with New Delhi, this does not represent the expansive desire of an omnicompetent state. Rather, the ability of the state to order its people and territory was repeatedly problematised by protests, riots and campaigns. To tackle this diversity of resistance, the Delhi Administration collected a formidable arsenal of disciplinary techniques, stretching from the violent to the visual and the juridical. This chapter will analyse the means used to discipline Delhi's urban space, paying attention to the identities, visibilities and technologies assumed by this regime of government. These assumptions will then be examined in practice during the major nationalist problematisations of public order in 1930–2 and 1942. Particular attention will be paid first to the inadequate conceptualisation of the law in Foucault's writings on discipline, given the central role of ordinances and legal proclamations in the colonial struggle against nationalist movements.

Sovereignty, law and discipline

Foucault's claim that sovereign forms of power had been subsumed within disciplinary and governmental forms, thus invalidating the theoretical apparatus of sovereignty, had important implications for his understanding of law. In his 1975–6 lectures, Foucault claimed that juridical thought in the West had emerged around royal power in the Middle Ages and was inextricably intertwined with the sovereign (Foucault, 1975–6 [2003]: 25). These claims were repeated in his later work that argued that law was pre-modern and detracted attention from the real functioning of power (Foucault, 1979: 88–9). As such, law was depicted as vulnerable to colonisation by the disciplines, it was dislocated from the sites of practicing power and it was portrayed as incompatible with modern forms of rule (Smith, 2000: 204).

Alan Hunt and Gary Wickham (1994) have clearly shown how problematic Foucault's linking of sovereignty to juridico-political discourse and the law is. This is demonstrated not least through Foucault's ambivalence over the role and function of law. The disciplines were at times claimed to be the dark underside of the law, where a carceral continuum naturalised legal power and legalised the technologies of discipline (Foucault, 1977: 222–3). This juridification of forms of discipline allowed law itself to become normative (Foucault, 1979: 144), yet Foucault continued to insist that law and discipline were incompatible.

Whether this process marks the colonisation of law by the disciplines or the uptake of the disciplines and their expert advisors by the judiciary when required, it is clear that modern society is pervaded by laws that enhance surveillance and discipline. Kevin Stenson (1998: 341) has gone further and argued that sovereignty remains influential not just as a theoretical approach, but also as a legal–political condition of possibility for liberal government. As such, law in the modern period functions in compatibility with new disciplinary and regulatory tactics, while also retaining some of its functions with regard to defence, territory and rights, whether of the central state or the democratised rights of the individual subject. Mark Neocleous (1996: 68) has suggested that this functionality is mediated not by the sovereign, or some abstract state, but by political administrations that bridge the state and civil society. Such a focus challenges the dissolution of the law in Foucault, showing how administrations create, are subject to, and act through, the law. As Foucault actually suggested, the legal imperative *is* normative, but the law codifies norms, while the techniques of normalisation associated with discipline and security develop from below the law or in resistance to it (Foucault, 1978b [2007]: 25 January).

In the colonies, as such, law was used in translated forms, without the liberal checks that existed in Europe. As Philip Howell (2004b) has shown in the case of the regulation of prostitution in nineteenth-century Gibraltar, the law was consistently used to regulate sex work and venereal diseases in a territory of overbearingly close governance by the sovereign state. Similarly, Sally Merry (2000) has argued that the law was a core institution of colonial control, serving commerce and capitalism by producing free labour and privatised land, while also serving as an ideological cornerstone of the civilising process. In accepting this, Nasser Hussain (2003: 102) has also shown that law was used to determine a state of emergency, and thus to appear at its own vanishing point, determining the rules of its failure. Just as sovereign forms of power co-existed with those of government, so also the norm existed alongside states of exception; the stately sovereign right to determine when the law was to be used to create an emergency. These emergencies were *not* outside the law, but a different logic of it. Hussain stressed that the norm and exception, law and emergency, were tied together in the colonial context as the government had to situate itself between perceived primitivism and oriental despotism. Colonial lawful rule was, thus:

> ... a median category. It is a form of sovereignty and governmentality: a rule that is lawful, as it lays claim to legitimacy through law, but also one that is literally full of law, full of rules that hierarchicalize, bureaucratize, mediate, and channel power.   (Hussain, 2003: 32)

Yet, within this abundance of law were periods of marked absence. Martial law simultaneously marked the highest form of law and the absence of law altogether. A historically and geographically, and thus racially, specific state of 'necessity' justified the declaration of martial law that exposed the laws' ambivalent relationship with violence. Since state violence had to be distinguished from that of the mob, increases in violence led to increases in the calls to regulate it. Violence and law were, thus, constantly forced into negotiation with their limits. These negotiations were not abstract or philosophical, but practical and situated in the context of local politics and the perceived threat to the state.

The anti-colonial nationalist movement in India posed serious threats to the ability of the state to maintain law and order. The mass movements of 1919, 1920–2, 1930–2 and 1942 presented the Delhi Administration with the need to declare an emergency. This did not, however, lead to Delhi resembling a camp populated by *homines sacri*, human beings without legal rights who could be sacrificed without penalty (Agamben, 1998; Gregory, 2004). Rather, technologies of discipline were warily deployed and developed, the evolution of which provides insights into the forceful imposition of order in the capital region.

The example of Delhi shows that the law was used to justify and legitimate acts of violence and martial force. Yet, it was also deployed to discipline space and restore order. In addition, it worked towards 'moral effects', mainly through fear, that would create a self-regulating subject as much as a docilely disciplined one. All the disciplinary effects presumed a certain subject, one that could not only be worked upon but was also inherently untrustworthy. It was upon this identity assumption that the city was viewed and ordered in times of crisis.

Identity: Civil Lines

The Durbar of 1911, at which King George V was lavishly received in Delhi and the capital transfer was announced, has been examined for its symbolic and political significance (Cohn, 1996). Despite this, the local machinery on which it depended has not been analysed. The influx of royalty, bureaucrats and social elites from all over the country meant that Delhi's local population had to be controlled and disciplined. The Delhi Durbar Police Act (1911) provided special and temporary arrangements for the administration of the Durbar camps, the surrounding areas, and Delhi Municipality.[41] It comprised a list of 33 activities, or 'modes of conduct' in terms of government, that were prohibited for threatening the 'public'. However, the elite 'public', which was the focus of attention here, would be residing mostly in the Durbar camp, occasionally in the Civil Lines, but rarely in the old city. As such, the prohibitions were differentiated into those that would be applied to protect the population of the Civil Lines and Durbar camp, namely Europeans and high-ranking Indians, and those that could feasibly be extended to the old city to improve its appearance for those passing through or residing there. The modes of conduct that were enforced in these two areas reflect the differential *identities* that were assumed and the settings that were required for either India's elite or its subaltern masses.

In defining forms of conduct that were not acceptable in the 'Civil Ward', the Police Act revealed the uncivilised manner in which it feared the Delhi population might conduct itself. The prohibitions can be classified into three broad bands that cohere around an orientalist stereotype of the colonial native: that they were closer to nature, more embodied and without self-regulation. Using these categories but maintaining the words of the Police Act, it was stated that imprisonment or fines would be levied on anyone within the Civil Ward who was believed to be:

(1) Closer to nature:
    (a) causes or permits any animal to stray or graze;
    (b) slaughters any animal or clean any carcass except in places approved therefore;
    (c) carries meat exposed to public view.
(2) Too embodied:
    (a) '[c]ommits nuisance by performing the offices of nature in other than the appointed places or wilfully and indecently exposes his person';

    (b)   bathes or washes either his clothes or person, or throws offensive matter, rubbish or pollution into the water supply;

    (c)   removes nightsoil without using the proper tools or fails to clean away spilled matter;

    (d)   '[l]oiters or begs, exposes or exhibits, with the object of exciting charity, any deformity or disease or any offensive sore or wound'.

(3)   Without self-regulation:

    (a)   '[s]olicits for prostitution or distributes notices or advertisements for prostitution, or, except within the limits of the Delhi municipality, establishes or keeps a resort for the purposes of prostitution or resides in a resort for the purpose of carrying on the trade of a prostitute';

    (b)   gambles;

    (c)   is found drunk and disorderly;

    (d)   cruelly beats any animal;

    (e)   fights or creates a disturbance.

Those prohibitions that were applied to the Civil Lines *and* the old city were less ambitious than the highly prescriptive mode of conduct outlined above. Their aim was not to protect the civilities of the occupants, but to guard the public space and political property of the municipality. The assumption was that although the normal subject could not be expected to govern himself or herself in a truly civilised manner, they could be disciplined into political obedience. Again, the Police Act stated that punishment awaited whosoever:

(1)   In public space

    (a)   drives, rides or leads a vehicle or animal in an area prohibited by the police or in a rash or negligent manner;

    (b)   neglects to take due care not to frighten horses while in the possession of a camel or elephant;

    (c)   leaves a vehicle without control;

    (d)   keeps an animal or vehicle standing for longer than necessary.

(2)   In political property

    (a)   trespasses on any parade ground or within the limit of any Camp;

    (b)   damages, breaks or throws down any direction-post, lamp, lamp-post, tree, bush, or other Government or Municipal property, or extinguishes any light in any public place;

    (c)   defaces or writes upon any notices, documents, building, monument, tent, post, wall, fence, tree 'or other thing';

    (d)   affixes any bill, notice or other document on any bill, tent etc.;

    (e)   *'Loiters, lurks, or is found in any place under such circumstances as to raise a suspicion that he was about to commit, or aid in the commission of, an offence, or that he was waiting for an opportunity to commit an offence*.[42]

The combination of the final stricture within the same list of 33 prohibitions as the dictates on the Civil Ward is telling. This unites the first assumptions on identity, that the Delhi's normal subjects could not govern themselves in a civilised manner, with the latter assumptions not only on their inability to properly conduct themselves in the public sphere but in their criminal potential and politically threatening nature. The classic liberal colonialist assumption was that the colonial population needed to be disciplined as well as governed (Valverde, 1996; Mehta, 1999; Hindess, 2001). This latter task would continue over the next 35 years and would be enforced through the visibilities and techne examined in the rest of this chapter, which aspired to the authoritarian scope and power outlined in the last criteria detailed above.

Visibility: Disciplined space

The attempted assassination of the Viceroy in 1912 prompted a reorganisation of security procedures regarding the protection of high-ranking officials. In 1915 a new set of 'Rules for the Protection of the Viceroy' were formulated, allocating responsibility to the head of police for each area the Viceroy travelled to.[43] Different guidelines dealt with the Viceroy in camp, station or private realm; when travelling by rail, steamer or road; and for public arrivals, departures and functions.

The last of these guidelines outlined a division of space in line with the 'art of distribution' outlined in Foucault's discussion of discipline. The 'Rules for the Protection of the Viceroy' can be seen as an early formation of a form of *visibility* that would recur throughout the colonial period, being more comprehensively combined in the 1930–40s to form 'diagrams' for disciplining the whole city. Without the cohesive rationality of a diagram, these rules appear more as temporary re-territorialisations, to use Deleuze's and Guatarri's (1987) terms. The smooth flowing of people over space would be canalised through 'striating' space and installing common measures to determine the similarities and differences within these lines of order (Patton 2000: 112). The rules stated that:

> [t]he route should be *divided* up into sections for purposes of police arrangements and there should ordinarily be a *small reserve allotted* to each section. Each section should be in charge of a responsible officer whose duty it will be to make himself acquainted with the geography of the houses on the route and of all converging side lanes, passages etc and to instruct his men in these details. He must make his dispositions in such a way that he can *communicate* with his reserve with the least possible delay.[44]

As such, the route would be fragmented and administered to the responsibility of a small police force, whose function it would be to make the route safe. Each unit would remain in contact with the responsible officer so that the collected information could be analysed and the risk of each fragment assessed. Throughout the rules, the emphasis remained on visibility. Narrow streets of less than 25 yards (23 m) were to be avoided while the route had to be restricted to 'respectable quarters'. When such areas could not be avoided, the houses themselves would be penetrated in order to warn the owners of their duties and to position plain-clothes officers on the roofs. Constables placed 15–20 paces apart would subdivide each busy street while plain-clothed officers would move within the crowd. Crowds that impaired visibility were to be diverted or the front row of onlookers made to sit down, while officers would patrol the back of the crowd. When at rest, the space around the Viceroy was divided into concentric rings of officers with a European inspector as his personal bodyguard.[45]

These arts of distribution were not just mobilised for the Viceroy. As a ceremonial space, Delhi played host to the most prestigious men and women in India, yet this was in an atmosphere of growing nationalist critique, whether non-cooperative or violent. As such, fleeting attempts to discipline space had to be made during high-profile visits, attempts that sought to make the landscape and its occupants visible and, hopefully, safe.

The visit of the Duke of Connaught to Delhi between 7 and 15 February 1921 saw the rules of protection being tailored specifically to the Delhi environment.[46] A conference held by the Home Member of the Secretariat on 24 December 1920 laid out the security operation in detail.[47] The Duke's route from the station to the Viceregal Lodge, still located in the temporary capital to the north, was lined with troops facing the road and police facing the crowd, while the distance between the carriage and the crowd was set to a minimum of 12–15 yards. Barriers were set in place to hold back the crowds at the potential trouble spots of Kashmeri Gate and Chandni Chowk.

For the ceremonial opening of the Memorial Arch and Legislative Assembly in New Delhi, 63 police staff guarded the route while 250 British troops were placed in garrison at five weak spots along the route. These included points of constriction, such as the Lothian Bridge, Kashmeri and Delhi Gates, and more open, vulnerable spaces, for example at King Edward Park or near Delhi Jail. In order to maintain the simulacra of a popular procession, it was suggested '[i]t could be arranged that these picquets were not too much in evidence at the moment when the Duke passed'.[48] There was, however, popular hostility to the amount of policing imposed on the city, although a questioner addressing the 'high handedness' of the police in the Legislative Assembly was told that there had been no official complaints regarding the more than 3,000 police staff on duty.[49]

While the disciplinary organisation of space would later be more rigorously ordered through comprehensive urban diagrams, the earlier plans for the city in the event of a riot relied upon deployments of the police with relatively little statistical or administrative visualisations of geographical risk or vulnerability. In 1889 Major Davies had stressed what a volatile city Delhi could be, capable of spreading rumours through vast numbers of people in a short amount of time (Parsons, 1926: 147). Despite this, the notes made to advise Delhi administrators in 1906 insisted that the city was easy to patrol as long as the strategic point of *Kazi Hauz* square was taken (Parsons, 1926: 162). The police were evenly distributed throughout the walled city, although with an over concentration of staff based at Sabzi Mandi in the north-west, which the 1906 notes themselves admitted was out of place.

By 1914 the police had more detailed plans although the specification of the measures to the Delhi landscape was still rudimentary. In the event of a riot, all officers and men reported to their allotted police station while some were sent to guard vulnerable sites such as the Bank of Bengal and the Treasury. It was up to the Superintendent of Police to '... take such action as seems desirable regarding further concentration of movement of men assembled in different stations'.[50] However, there was a shift towards spatial demarcations by means of a memorandum attached to the rules. It stated that two essential points must be held in the city. The first, Sadar Bazar crossing, prevented groups from gaining access to the walled city from the west, while 'Kashmere' and Mori Gates in the north were to be held as they protected the Civil Lines.

Although the police were able to impose a disciplinary ordering of space in the city with a preordained itinerary, they lacked the means of visualising the city during a major outbreak of violence. The next revisualisation of the city came with the problematisation of the early 1930s during the Civil Disobedience movements. But there had also been developments in other fields. In 1914 the police were ordered to collect muskets and 50 rounds of buckshot, with the remaining men taking lathis and swords, while the sergeants would be mounted on horseback. The 1920s were a period of rapid development in terms of the means and methods through which disciplinary space could be imposed.

Techne of discipline: Towards a 'moral effect'

The techne of government attends both to the moral guidance of conduct and to the disciplinary maximisation of aptitudes and the optimisation of *performance* (Dean, 1996: 61). This section seeks to address the governmental technology in colonial Delhi that was united by the strategic rationality of attempts to instil discipline and political docility. However, in practice, such attempts were necessarily embroiled with attempts to regulate conduct, although this was often less from a distance and more entwined with the politics of fear. A series of developments in the 1920s enabled the police to extend a degree of disciplinary force over the city in times of need. However, Nandini Gooptu (2001: 115) has shown that throughout northern India in this period there was an ongoing tension between anti-colonial and communal 'emergencies' and the 'economy' drive of the Government during the Great Depression era of the 1930s. The tension between cost and need was also evident in the urban politics of Delhi.

Technology

The Rowlatt Disturbances of 1919 set in motion a debate about what techniques of government the state should have at its disposal. While the army remained accessible, with bases in the Fort and the Delhi Cantonment, the disciplinary technology to which the civil administration had access underwent a programmatic review. Within 4 months of the Rowlatt disturbances, the General Officer Commanding the Delhi Brigade sent a letter to the Chief Commissioner requesting the formation of an *Armoured Motor Battery*. This would consist of three armoured Government cars with machine guns and three soldier drivers to maintain and utilise them. A unit of armoured cars had, in fact, been supplied to Delhi in 1916, consisting of three cars each requiring two drivers and three other members.[51] However, when the cars arrived in Delhi in December 1916 their tyres were faulty, they had broken down on the way and were deemed by Brigadier General, Earl of Radnor, to be completely useless and a danger in operation.

The Rowlatt disturbances of 1919 sufficiently alarmed the local army Commanding Officer such that he reasserted the case for a (functioning) armoured car battery. He argued, on 28 July 1919, that 'the location of this battery at Delhi would be a most powerful asset to the cause of law and order in the event of disturbances, and one would hope that the European and Anglo-Indian residents of Delhi would be only too keen to join up'.[52] Stoler and Cooper (1997) have stressed that colonial society was one with its own internal ruptures. One tension that repeatedly surfaced was that between the administrative bureaucracy and the military authorities. Although an often-successful duo, this was a marriage of convenience, each at times relying on the other, and at times prone to disagreement and bickering. Conscious of the need for an efficient disciplinary apparatus, Mr C.A. Barron, the Chief Commissioner, was aware that without proper training and use, the armoured cars could easily have incited more mob violence than they subdued. Barron replied to the army officer on 21 August 1919 that without trained men, mounted machine guns would have 'been more a danger than a strength'[53] in the Rowlatt troubles. However, by 21 May 1921, sufficient funds had been allocated and an Armoured Motor Battery was (re)-established as part of the Delhi Movable Column.[54]

Armoured cars were used extensively during the Civil Disobedience campaigns in Delhi, as will be shown later in this chapter. However, the extent of this use led to new instructions for the 'employment of armoured cars and light tanks in aid of the civil power' being issued in 1935.[55] The instructions were as much about the theatricality of urban discipline than they are about military strategy. The cars were agents of sovereign power, seeking to instil self-conduct through fear, as well as machines to marshal politically docile subjects. They displayed the fragile constitution of sovereignty by affect that drew on a logic that did not concern innocence or guilt, or even retribution or rehabilitation; rather it was '... purely performative, the purpose of which is the sheer manifestation of power itself' (Hussain, 2003: 124). Central to this function was the 'moral effect' of tanks and armoured cars, whether being used as escorts, for placatory patrolling, dispersing assemblies or preventative action. For the latter, the armoured cars were given areas of a city to patrol, but only the main streets. At no point were they to come into regular contact with the crowds, as the 1935 instructions dictated:

> The guiding principle to be observed is that patrolling should not be carried out too frequently or the populace will become accustomed to the machines and their moral effects will be reduced. It will be best to carry out a patrol immediately on arrival in the affected area, in strength, and then to withdraw the armoured cars to a locality screened from the public view. Rumour will then tend greatly to exaggerate their numbers, and thus increase their moral effect.[56]

Similarly, the moral effect of vehicles was said to be greatly reduced by unreliable machinery, especially 'if crews are seen to dismount and attend to engines'. Besides this, the cars were acknowledged to have many limitations: their commander could not contact them, the range of vision was limited, they could be blocked by barricades and they were vulnerable to attack by crowds, in which case the only response was said to be to open machine gun fire. As such, the machines were urged to be used with caution and in collaboration with the host of other measures that had been developed to control the towns.

The technological debate had progressed a step further when, on 30 April 1922, a Secret Memorandum was distributed detailing the instructions for the use of *aircraft* in the aid of civil power.[57] The use of aircraft against rioters was not within the King's regulations of the Royal Air Force. These stated that they should be used only for reconnaissance or against the illegal use of aircraft outside of times of martial law or state of war. However, concessions were made in cases where life and property were rendered exceptionally insecure due to riots that encouraged 'racial feelings'. This is a clear example of the way in which the 'necessity' of exceptionally violent technologies of control was defined through the racially specific logic of colonial governmentality. In cases where the military could not offer aid, then aircraft assaults, although posing 'a huge risk of firing on the wrong people' and being inaccurate even if they were fired, were allowed. Having gained permission and identified the right crowd, the procedure was to warn them, by light or horn, to fire a few rounds if no response was met, then to use machine gun fire and finally to drop bombs as a last resort, although never on towns.

Although aircraft were never used aggressively in Delhi, *tear gas* would prove to be a valuable addition to the colonial arsenal. Just as there was a differential desire for certain techniques between local Government and Army officials, so there were inequalities between the local and central Government. This debate itself was prompted by an international development in urban disciplinary technology. In 1928 an article appeared in the *Journal of the United Service Institution of India* detailing the Mob Street Platoon in Shanghai that used tear gas to suppress riots (Radley, 1928). The article was forwarded from the Deputy Secretary of the Home Department to the Chief Commissioner in the belief that the motor lorry and, especially, the '... moral effect produced by it on crowds ...' would be of great use to local governments.[58] In contrast to the buckshot and swords of 1914, the sheer range of technologies on offer was astounding: a sawn-off shot gun, a Thompson sub-machine gun, a MC Bride Tear Gas Gun, tear gas hand grenades, smoke candles, gas masks, hand cuffs and crow bars. The theatricality of this arsenal was also stressed. The presence of the unit was hoped to be a sufficient deterrent, with intimidation increased by a special bus or lorry 'or a terrifying aspect' and tightly disciplined troops: 'The moral effect on the mob, caused by seeing the sections reform, is greater than that caused by the actual charge' (Radley, 1928: 33).

The proposals met with resistance from within the Delhi Administration. The Chief Commissioner, A.M. Stow, wryly pointed out that the primary concern was that of who would pay for the platoon, while the Deputy Commissioner stressed that Lachrymatory gas would hang for days in Delhi's twisting lanes and be more of a punishment to the people in the surrounding houses or the police themselves. However, with a change of personnel came a change of approach. On 19 February 1930, Delhi's new Chief Commissioner, J.P. Thompson, sent a letter to H.G. Haig, the Home Secretary. Thompson was chasing up the Shanghai circular and urged Haig to consider the use of gas for mobs 'incited by revolutionaries'. He continued:

We want something as effective as fire action in its immediate object, but less serious, for all concerned, in its after effects, something which will not kill, will give no crown of martyrdom and yet reach a larger proportion of the offending crowd and will not lead to much bitter recrimination afterwards.[59]

Haig replied that a Police Conference in January 1929 had decided against the use of gas during civilian agitation.[60] However, by January 1936, the Government was carrying out trials of gas, one of which was attended by E.W. Wace, the Senior Superintendent of Police for Delhi. The use of gas was demonstrated against both violent and passive crowds and for use in enclosed spaces, in broad roads, village squares or open spaces. Wace reported back excitedly that, '[t]he moral effect on a crowd of the discharge of gas hand grenades and of the discharge of gas blast cartridges and shells from a riot gun would be very considerable, and I consider that a ricocheting gas shell discharged from a riot gun against a crowd and exploding in front of it would in itself cause an ordinary crowd to disperse.'[61] While American-invented 'gas motor cars' that dispersed gas from cylinders as they drove through the crowd were not adopted, by 1937, Army Headquarters was advising provincial governments that tear gas could be used by the police.[62]

The Chief Commissioner, J.N.G. Jenkins, sought the advice of the Home Department, who outlined the circumstances that justified the use of gas: if the occasion demanded it, if it could be justified on the 'grounds of humanity' and if it would be successful. Despite experiments showing that entirely satisfactory products were not yet available, by September 1938, Jenkins had been convinced by the success of tear gas in the U.S.A. arguing that it was more humane than physical violence.[63] He was outraged, however, at the cost and suggested the central government pay for the training and provision of materials. They refused, arguing that Delhi only needed half the men suggested and that savings could be made in the existing budget. Jenkin's retorted, stressing Delhi's 'special circumstances', that were in no small part due to the government, and asked whether the government would pay and he should proceed, or whether the plans be abandoned? The Government backed down and the men were sent for training at Phillaur. Within a few years the tear gas unit was used to great effect during anti-colonial and communal disturbances.

Military support

Despite the will being shown to equip itself with the finest machinery, the army was reluctant to use its force in civil issues. As Nasser Hussein (2003) has stressed, a state of emergency that would justify the use of the military as against the police depended on the definition of 'necessity' and this was a definition that was passionately debated within the government. Radley's (1928) article on Shanghai had stressed just how much the army regarded mob dispersal with distaste because it was 'not soldiering' and because they were bound to do wrong in the eyes of others. Upon similar lines, tension between the bureaucratic and militaristic arms of government had been exposed after the extensive nationwide use of the military in 1919. The Secretary of the Army sent a memorandum to all Chief Secretaries and Chief Commissioners on 19 October 1920.[64] It was shown that the principle of the police acting without the military had been repeatedly affirmed since Sir Ashley Eden's classic statement of 1879. This acknowledged that the 'maintenance of tranquillity and the safety of the British Government in India' depended on the military forces, yet insisted that the protection of order was the duty of the civil government.

On 26 May 1921, Delhi's Chief Commissioner responded with an equally forceful statement of the ethos that would underwrite the approach to violent government in Delhi for the next 25 years. Barron questioned the dividing line between 'local disturbances', violent disturbances, rebellion and a 'rising against the government' that the Army had suggested. As with the biopolitical governance of space outlined in the Improvement Trust debates in the following chapter, the disciplinary government of Old Delhi was a special case, as the Chief Commissioner insisted:

> Moreover in a place of such political and strategic importance as the official capital of India it is necessary to be prepared to use even overwhelming force to nip any incipient trouble in the bud. The consequences of failure to do so would be too serious to be contemplated by a responsible Government. Fortunately Delhi, as a defensive post with an obligatory garrison, and the base of a movable column in the Internal Security Scheme, is never bereft of a military force sufficient to deal with all likely eventualities.[65]

The combination of military and civil measures conformed to the Punjab Internal Security Instructions (1922) that were consulted in the following debates.[66] These stated that civil measures were to include quelling local temporary disturbances and scattering hostile crowds that threatened Europeans and loyal Indians, protecting lines of communication and safe guarding 'treasure' and records. If events escalated, then military measures could include announcing Martial Law, instigating action by movable army columns and collecting Europeans and loyal Indians into stations to protect them.

In 1928, the Army Headquarters attempted to reduce the military personnel stationed at Delhi, to which the Colonel Commandant and Chief Commissioner responded with a lengthy defence of the need for a strong military presence in Delhi. The argument centred around the need to protect the capital and the inability of the local police force to do this in times of crisis. While the troops in Delhi were retained, the Adjutant General issued a report on military aid to civil powers in 1932.[67] While the official Internal Security Scheme was well planned for, it was rarely used; yet troops were called out to aid civilian powers on an increasingly regular basis and needed clarification on certain procedures. Outside of periods of martial law, the troops were used for preventative work, active suppression and the retention of peace after an outbreak. Yet, their role was to be strictly delimited to forestalling or ending violent outbreaks, at no point taking on peace-keeping duties.

By 1942, these distinctions had crystallised into 10 separate schemes that were divided into those with a military or civilian focus that could be jointly brought into action as the specific situation dictated. These had been, and were, used in conjunction during the mass movements to re-assert the authority of the state, as will be shown at the end of this section.

Juridical

While the Rowlatt disturbances had demonstrated the strengths and weaknesses of Delhi's armed and police forces, it had also demonstrated how their powers could be augmented by legal means. After 2 weeks of troubles, the Chief Commissioner issued the Prevention of Seditious Meetings Act (1911) on 17 April 1919.[68] This banned unauthorised meetings in public areas and, as the Chief Commissioner commented on 11 October 1919: 'The effect of the notification at the time of its issue was instantaneous in producing a sense of peace and restoration of order in Delhi City by the mere prohibition of unauthorised public meetings, at which violent speeches grossly misrepresenting the motives and actions of Government could be delivered.'[69] The act remained in use until February 1921 but was not accepted passively by public commentators. When the Duke of Connaught officially opened the Legislative Assembly during his visit in 1921, *The Tribune* newspaper pointed out that the city in which the ceremony took place still did not have the right of free speech.[70] This was just one of the many disciplinary deployments of the law. These constructions of temporary spaces of law were not simply negative or extractive; they were about the ordering of bodies on the ground and installed an effective mode of surveillance when they were in use.

By the outbreak of Civil Disobedience in March 1930, the Government had arranged an array of preventative legal measures that it could use to suppress 'disorder'. Many of these laws were formed in other parts of India but could be extended at times of need. While some of these powers attest to a sovereign form of power that forbade and disabled, they also contained powers to control the movement of specific individuals. These gave the Delhi authorities the power to

(1) detain suspects under restraint without trial;[71]
(2) place suspects of terrorism under supervision, restricted movement or to suspend them in jail;[72]
(3) place people under bonds of good behaviour;[73]
(4) force persons to abstain from a specified act;[74]

(5)   force persons to forfeit seditious literature;[75]

(6)   announce an association to be unlawful and membership of it to be an offence.[76]

During Civil Disobedience (1930–2), a further host of repressive legal measures were created that allowed the local administration to

(1)   force publishers to pay deposits for return on good behaviour;[77]

(2)   forbid the picketing and boycotting of government servants;[78]

(3)   outlaw acts that undermined the Administration;[79]

(4)   forbid unauthorised newssheets and newspapers;[80]

(5)   control movable and immovable property, persons and public services.[81]

Though granting extensive powers, these laws were imperfectly implemented and usually contested. Their usage was time and place specific and relied as much on the whims and fancies of their implementer as much as the actual need for their implementation, as will be practically demonstrated in the case studies of the Civil Disobedience and Quit India campaigns.

Dispositions

While imperial governments assumed a *global* geography of liberty, these distinctions also structured *local* projects and programs. While using legal, military and policing technologies of government, implementations of disciplinary power also relied on place-bound dispositions of memory and knowledge. Delhi's disciplinary techne was ingrained with the ethos that dictated the protection of the capital at all costs. This was displayed in January 1930 when the Delhi Administration performed an Internal Security Exercise in which a mock run-through of a rebellious uprising was performed to monitor how the local forces would respond.[82] These imaginary actions indicate how agitation was expected to occur and how the local geography of risk would inform the reaction of the armed forces.

The imaginary uprising began at 10 AM on 14 January 1930 with 'fanatical Mohammedans' gathering at the Jama and Fatehpuri Masjids, touring the streets and beating some Europeans to death. Hindu and Sikh seditionists fanned the flames while revolutionary arms were distributed. By 12 PM the situation had rapidly deteriorated and a mob was soon expected to head north, to the Civil Lines, or south, to New Delhi. The ability of the police to deal with such an assault was questioned, although they managed to defend the Civil Lines from the imaginary mob at 10 AM the next day. A military force was detached as a garrison to protect the functioning of the Government of India in New Delhi, while the impossibility of outwitting the revolutionaries in Old Delhi was admitted because they worked 'on internal lines of communication'. Displaying Beadon's enthusiasm to use 'overwhelming force to nip incipient trouble in the bud', the report terminated with one final statement that embodied the ethos that would characterise the approaches to nationalist uprisings in Delhi throughout the 1930s and 1940s. As part of the capital region, Old Delhi had to be disciplined at any cost:

> Will need reinforcements, but cannot afford to lose control of Delhi as the new capital. Old Delhi has great food supply and resources, cannot afford to wait. Must strike with every man and weapon available.[83]

The problematisations discussed in Chapter 2 occurred in governmental channels utilising acknowledged enunciative modalities. The Delhi Improvement Trust (DIT), as the next chapter illustrates, was challenged by petition, critique and protest. Yet, the intensity of force associated with the disciplinary apparatus provoked much more obvious acts of problematisations that brought about new, and violent, articulations of the disciplinary techne in the old city.

Problematisations

Civil Disobedience 1930

On 5 April 1930, *Mahatma* (Great Soul) Gandhi arrived at the Gujarati coastline and proceeded to make salt from the seawater, thus breaking the British monopoly on salt manufacture and marking the beginning of the first Civil Disobedience campaign. The initial reaction of the Deputy Commissioner was to hold the more repressive apparatuses back, in an attempt to tackle the nationalists on their own non-violent terms. The police were initially forbidden from taking lathis (staffs) to protest meetings in order to prevent provocation.[84] However, the ensuing disturbances convinced the authorities to employ more severe tactics. Armed troops were dispatched to further meetings while mass arrests of Delhi's political elite were ordered. Despite the increasingly harsh reprisals employed by the Delhi Administration, it would be a mistake to portray the local apparatus as a lumbering, and solely punitive, machine. While the policing policy was certainly *repressive*, in that it stemmed the increasing flow of people willing to reclaim Delhi's public space, it was also self-consciously *productive*, recognising that popular protest was not the manifestation of some form of collective subconscious, but was a phenomena contingent upon local circumstances. The Chief Commissioner refused to ban meetings or processions, commenting on 4 May 1930 that

> [t]hey act as a safety valve at a time of such excitement as the present and to have prohibited them would have been to precipitate an immediate and unnecessary

crisis – so far as we here in Delhi are concerned. Moreover there is no question that up to date police definitely control the city to every and any degree which is essential.[85]

Despite this approach, the Chief Commissioner insisted in the same statement that '... at any time we are prepared to take very definite and drastic action to preserve law and order and to enforce the authority of Government if such action is forced upon us.'[86] In preparation for such an occasion, an Army had been on standby at the Fort since 10 April. The next day, after Europeans were spat at in the street, the armoured car patrol was established from Kashmeri Gate in the north of the city, to the jail south of Delhi Gate, while the more substantial troops of the Delhi Independent Brigade Area were put on standby throughout the next two days.[87] The armoured cars were used again after the arrest of Gandhi on 5 May, and helped to re-establish order between Chandni Chowk and Lahore Gate the next day following the firing by the police on crowds outside the Gurdwara Sisganj (see Legg, 2005b). The armoured cars had the desired effect, the very appearance of which was, as the Chief Commissioner phrased it on 8 May 1930, 'peculiarly prophylactic'.[88]

In terms of legal techniques of discipline, the Home Secretary suggested that the Chief Commissioner might introduce the Prevention of Intimidation Ordinance into Delhi Province. This prompted a debate about the need for such disciplinary measures in the face of non-violent protests such as the picketing of foreign cloth and liquor shops. Delhi's Public Prosecutor stated on 21 April that '... it is perfectly clear that the system of picketing is non-violent and peaceful. Picketing is really nothing but free persuasion' and lodged his complaint that the law would make non-violent persuasion an offence.[89] The Deputy Commissioner also commented on 23 April that if 'molestation' were defined as 'the attempt to stop someone doing or not doing an act, or loitering near a place where someone works or resides' the law would effectively ban people from standing around in public.[90] Irrespective of these protests, the Ordinance was made available on 30 May 1930, just after the imposition of the Press Ordinance.[91] Following the shootings and disturbances at the start of May, Section 144 of the CrPC was declared for a month and then extended for another 30 days.[92] This law, under which people had to abstain from specific acts, forbade people to:

(1) carry any firearms, dang, lathi or other weapon in any street or public place within the Delhi Municipality, New Delhi Municipality, Civil Lines, Notified Area and Fort Notified Area;
(2) collect or keep any of these weapons or any heap of stones, bricks or any other missile in or on any building with the same limits, providing

that this order shall not apply to any Government Servant acting in the discharge of his duties;

This last component highlighted not only the illiberal nature of these disciplinary actions but also that, as Hussein (2003) pointed out, the state and its challengers often shared means of violence. As Mehta (1999) and Valverde (1996) stressed, this was not a contradiction in the system of colonial government but a by-product of the belief that illiberal techniques could be used to discipline those who had not inculcated self-regulatory liberal habits. The actions that were forbidden were exactly those that would be used in an attempt to enforce political docility. Other laws were introduced to stamp out the movement as it entered its fifth month. The Unlawful Instigation Ordinance was introduced in late July to combat a no-tax campaign, while Section 32 of the Police Act was used against liquor shop picketing, leading to the arrest of 176 people within a fortnight.[93] The severity of the combined technology of discipline meant that the movement caused little major discontent to the government until the Gandhi–Irwin Pact on 5 March 1931, following the inauguration of New Delhi.

Civil Disobedience 1932

Following the failure of the Round Table Conference in London in 1931, Gandhi was arrested on his return to India on 4 January 1932. The issuing of the Emergency Powers Ordinance on the same day gave the Government unprecedented powers of search, seizure and arrest. The lessons of 1930 had been well learnt. Non-violent non-cooperation's internal logic ran along the same lines as Foucault's conception of capillary power; once the state was denied its myth of centralised power, its authority fell apart during face-to-face contact. To avoid open confrontation, the Government used legal means to disorganise Congress and arrest the most dangerous politicians.

The Criminal Law Amendment Act (1908) was used to declare Congress institutions illegal leading to the confiscation of their officers and arrest of their leaders, to be replaced by unofficial 'dictators', who fared little better. By 10 January the four Ordinances that dealt with Unlawful Instigation and Association, Suspected Persons, Emergency Powers and Molestation, and Boycotting had all been applied to Delhi.[94] In the event of a major procession, CrPC 144 was used to arrest any men or women who tried to address the crowds while those suspected of political activity had their houses searched.[95]

Besides the devastating effectiveness of the technologies adopted, the 1932 Civil Disobedience campaign in Delhi marked one new development. The use of female protestors had proven deeply unsettling for the colonial authorities. Violence perpetrated against women exposed the illiberal and unmanly basis of an empire supposedly erected on liberal and gentlemanly values. On 7 January, the Central Government had written to the local authorities stating that the problem of dealing with women had not been solved in 1930 and that force should not be used against women in the future.[96] Maintaining his stance as demonstrated in his refusal to ban processions, Chief Commissioner Johnson recognised the use of force as counter-productive. He suggested that '... what one wants is a method which will bring moral rather than physical force to bear'. Recognising the moral persuasiveness of the nationalist leaders, the suggestion of sending 'respectable' women to bring pressure to bear on the families of female nationalist workers was rejected.

The use of female police officers of untouchable status saw the Delhi Administration challenging the nationalists on their own turf. Not only would the image of the cooperative Indian female replace that of the violent colonial male, it would also challenge the Congress supporters' claims to have surpassed caste prejudice. In the early 1930s many of the female activists that encouraged other women to participate were from educated, wealthy, high-caste families to whom such intercaste pollution would hitherto have been anathema. In addition, many Muslim shopkeepers who had suffered from Congress picketing welcomed the employment of more effective police figures.

In a Home Department memorandum issued on 23 June 1930, advice had been issued to local authorities on confiscating Congress buildings. It was advised that the '... effect would be increased if buildings were used for Government purposes, such as the accommodation of troops or police.'[97] In line with this advice on re-signifying the urban political landscape, the confiscated Congress Office on Chandni Chowk was used to accommodate the female police officers. By 25 January, the force comprised one Anglo-Indian head constable, eight Indian Christian constables and two low-caste Hindu constables.[98]

Despite having to be escorted at all times by male officers, the Senior Superintendent of Police reported on 17 February 1932 that he was pleased with the female police force and that 'the women of good social standing who have been accustomed to more than tolerance and courteous "handling" by the men police do not relish being handled by their own sex – some of whom are low caste.'[99] They were used for arrests, railway station patrols and transferring women between jails and led to a sharp fall in complaints against police violence towards women.

The powers amassed by the Government proved to be highly effective in disorganising the Civil Disobedience movements and while it attempted to function underground for the following few years, it had been effectively disabled. Following the Government of India Act in 1935, Congress sought election and did not engage in further agitation until the protests surrounding the War in 1939.

Quit India 1942

Following the 1939 resignation of Congress from the electoral posts it had accepted during the 1930s, the Quit India movement marked the nationalist's protest against India's unsolicited participation in the Second World War. Congress announced its intention to call for an end to British rule on 8 August 1942, leading to the arrest of almost the entire nationalist elite. The following movement was the largest mass mobilisation of the nationalist era and allowed the Government to call forth the full capacity of its disciplinary might. On 9 August, D. Kilburn, the Senior Superintendent of Police, outlined the Government's planned response to the Congress call for agitation.[100] The Delhi Administration anticipated 14 main forms of protest and listed at least one planned response to each form. The anticipated protests can be divided into three broad and overlapping categories; three concerned information, four dealt with the instigation of strikes, while five concerned specific sites. In response the Government recommended three main forms of action: the use of patrols, special ordinances, and pickets.

Regarding forms of agitation linked to *information*, one concern was with false telephone calls giving misleading information about rioting, thus each tip-off had to be corroborated using different sources of information. The remaining source of concern regarding information was that whispering campaigns would spread alarm or that rumours would be used to start a communal riot. While guilty individuals were to be charged if caught, the set response was to establish patrols throughout the city along the guidelines set out in the Communal Riot Scheme (CRS). This scheme will be examined in detail in the next chapter but it basically comprised an urban diagram of disciplinary order. The city was divided into smaller fragments into which police reserves were allotted and linked through police *patrols* to an informational base at the city's police headquarters. By 10 August, Chandni Chowk was being patrolled by 60 constables, 4 head constables and 4 Superintendents while forces were also allotted to other potential gathering sites in the city (see Figure 3.2). The duties of the patrols were made perfectly clear in the 'Scheme for Police Patrols' as outlined on 12 August 1942:

(1) to fire without warning on any persons committing arson or violent assaults including stone throwing;
(2) if organised demonstrations are encountered, to give a warning to disperse immediately and then to fire, provided the crowd is not too large;
(3) if the crowd is large, report to the police station.[101]

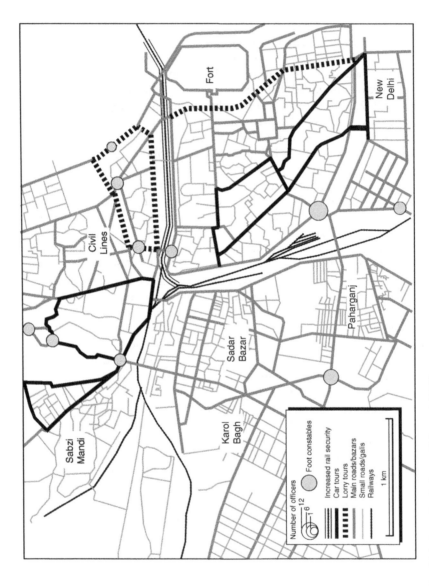

Figure 3.2 Pickets and patrols on 12 August 1942

The patrols were not to stick to fixed routes, but to cover the areas summarised in Figure 3.2. Lorries were used in 'more dangerous areas', being the city and suburbs, with 1 head constable, 10 foot constables and greater protection. These patrols aimed to prevent not only direct political violence but also interference with telegraphic, telephone and electrical supply wires, another of the main agitation forms outlined by Kilburn. The City Magistrate acknowledged the calming effect of these patrols on 18 August, although they were often too slow and heavy handed to deal with the mobile and easily dispersible local protests. Instead, magistrates and response forces were located at vulnerable areas, able to be deployed immediately, including the Kotwali, Maidens Hotel in the Civil Lines and Delhi Cloth Mills.[102]

The second main form of anticipated political action regarded the incitement to disaffection and to *strike*. This could take the form of a general hartal in which trade throughout whole communities would be suspended, or it could be targeted at specific sectors. The prime candidates included Delhi's major organisations of national importance, such as the Central Ordnance Depot or the Cloth Mills, it's public utilities or the vast number of police and troops stationed in the city. All three of these strike targets would be protected by police *pickets*. Rather than the dispersed reserves of the CRS, these would specifically protect loyal employees returning to work or prevent unauthorised persons gaining access to sensitive areas. In this function pickets would also be used to combat the third main form of expected agitation.

Following the experiences of the 1930s, the local administration was aware that the nationalists would target specific *sites*, the third anticipated form of protest. Action was anticipated at the District Courts, at the houses of high officials and at local grain dealers. Both pickets and patrols would be used to protect these sites, but for the targeting of high officials Kilburn recommended the use of a range of special *ordinances*. The Criminal Law Amendment Act or Police Act were both recommended, the Defence of India Act was suggested for dealing with whispering campaigns and attempts to arouse disaffection within the police, while CrPC 144 could be used against processions of over five people. Yet, the technology of government was pervaded by an *ethos* that dictated a certain distribution of power relations and energies. Perhaps the strongest combination of forces was rallied to protect the most valuable site of all. As shown in the first section, it is impossible to consider the policing of Old Delhi without the overriding principle that New Delhi would be protected at all costs. While the two cities had very different forms of policing, patrolling and picketing over radically opposed urban forms, they were united by forms of government and discipline that prioritised the European elite. The Quit India movement was put down with incredible efficiency. While symbolic and community protests continued, the disciplinary mechanisms had become incredibly effective at imposing an unsustainable but rigorous form of urban order, as the following account from Delhi in August 1942 suggests:

> This morning I woke at the famous old Cecil Hotel to find the entire hotel ringed about with guards (and an armed convoy drawn up to escort British officials to the office buildings in New Delhi. The convoy swung out into the street in military formation, evenly spaced with the regulation distances between cars, preceded and followed by a truckload of troops carrying fixed bayonets). We drove through Delhi and passed troops in almost all strategic corners. ... I saw (more) Tommy guns, rifles (assorted hardware) lying ready in the streets of Delhi (than I have seen in the past two months in China). ... The old city was completely under control, and military were visible everywhere. The streets had been cleared, the soldiers were ready and there was silence everywhere. I scoured the city thoroughly for scenes of disturbance – but the central city was pacified ...   (quoted in Mathur, 1979: 79–80)

At first glance, the disciplinary system of colonial governmentality in Delhi seems all powerful and all knowledgeable. However, three points are essential to bear in mind to refute this assertion. First, the above examples mark three major problematisations of colonial rule during which willingly mobilised people rejected the colonial regime of government. Second, the government failed to maintain its system of power that relied on discipline and conduct, not violence and contact. In 1942, between 11 and 12 August alone, the police opened fire on 20 occasions, releasing over 153 rounds of ammunition that killed 5 people and wounded 14. In the same period the army fired on 27 occasions using 266 rounds of ammunition, killing 14 and wounding 31 people.[103] Finally, the Government operated, as it always had done, primarily in the public sphere. Its priority was to restore order in the streets and trading places, driving dissent and resistance underground.

There was, however, an emphasis on encouraging compliance rather than forcing direct acquiescence. The was a need for a 'moral effect', whether through the dramatic impact of tear gas, armoured vehicles or female police, or through the considerate and human policing of festivals or the granting of processions. But behind the urge for a moral effect was not necessarily a humanitarian morality. Non-violent methods were less politically dangerous or expensive. Yet, it was within the costly and politically disastrous management of communal tensions in Delhi that the disciplinary apparatus achieved its most sophisticated realisation, as demonstrated in the following section.

'Religious Nationalism' and Urban Diagrams

Diagrams, communalism and policing festivals

Any discussion of the interaction between the heterogeneous groupings referred to as 'colonial' and 'nationalist' in twentieth-century India must also take account of 'communal' tension. This antagonism existed between the religious communities of, at the very least, Hindus and Sikhs on the one hand and Muslims on the other hand. The rise of violent attacks and social distancing between the groups cannot be considered outside of the colonial context. While this contextualisation was constitutional and socio-cultural, it was also explicitly political and practical. The police were involved in policing communal clashes and the annual festivals they often centred around. The evolution of police plans for these festivals marked the tethering together of modes of seeing the city, identity assumptions about the city dwellers, and technologies of urban discipline. In anticipating annual festivals incredibly detailed plans were constructed by which the police aimed to exert their authority over the city. These plans came together in the 1930s Communal Riot Scheme (CRS), which became the generic diagram of urban discipline that informed the later policing of annual festivals. Before exploring the phenomenon of communalism, the theoretical grounding for exploring these plans will be explained.

Diagrams

In *The Birth of the Clinic*, Foucault (1973) used three levels of spatialisation to explore the advance of modern medical knowledge in eighteenth to nineteenth-century Europe. Primary spatialisation looked at the way in which diseases were organised in tables and abstract knowledge on the page. Secondary spatialisations looked at the localisation of diseases in bodies and geographical masses. Tertiary spatialisations looked at the institutionalisation of the sick who were demarcated and treated in physical space (Philo, 2000). The coherence across these spatialisations emphasised how difficult it is to separate the representation of thought, the embodiment of ideas and their geographical dispersion and production. This flow of continuity was also addressed in Foucault's work on discipline, where the canalisation and channelling into certain 'diagrams' of these flows was examined.

Gilles Deleuze has suggested some provocative ways of reading Foucault in order to abstract what Donelly (1992: 200) has referred to as the 'epochal' elements of biopower, those programmes and political technologies that endure application beyond their specific context. Deleuze (1988: 33) suggested that in an abstract relationship to disciplinary material forms there exist 'diagrams' or 'abstract machines' that organise visible matter and impose conduct on human multiplicities. These diagrams form bridgeheads between the general arts of discipline and specific strategies and programmes.

The concept of the diagram allows disciplinary tactics to be thought of in relation to the urban form more systematically while also detracting attention away from the Panopticon as the model of discipline. This is especially pertinent given that Foucault drew on other articulations of disciplinary space in his work. These included the military camp and the plague town (Elden, 2001: 145), the former of which required acute observation of its distributed troops. As such, '... the camp is the diagram of a power that acts by means of general visibility', an 'ideal model' (Foucault, 1977: 171) of hierarchical observation. The plague town represented for Foucault '... a segmented, immobile, frozen space. Each individual is fixed in his [sic] place. And, if he moves, he does so at the risk of his life, contagion or punishment' (Foucault, 1977: 195). He continued,

> [t]his enclosed, segmented space, observed at every point, in which the individuals are inserted in a fixed place, in which the slightest movements are supervised, in which all events are recorded ... all this constitutes a compact model of the disciplinary mechanism.   (Foucault, 1977: 197)

The plague town and military camp, in one sense, allowed the fulfilment of a political dream, one of societal divisions and the assignation of place and identity (also see Jones, 2000: 47). Yet, as Ransom (1997: 44) has suggested, the people who formulated these diagrams were not authors or theorists but were more often administrators and police chiefs. These were people whose dreams very often saw the light of day, albeit in forms adapted to specific urban environments.

Osborne and Rose (1999) have used the term 'diagram' to refer to ways in which government, as the authoritative regulation of conduct, has been territorialised in an urban form. These diagrams are functional and technical rather than cognitive and can consist of drawings, plans, stories or programmes, the applications of which involve creativity on the part of those involved. Yet, Osborne's and Rose's emphasis on 'abstract cities' has been criticised for downplaying structural factors affecting the diagrams, such as class or community, and solidifying diagrams that are in fact maps of force with all the struggle that image suggests (Isin, 1998: 38). Colonial diagrams of disciplinary order were heavily structured by raced and classed notions of order, and were constantly challenged by different cultural formations and outright acts of resistance. The evolution of communal policing in Delhi also shows that the colonial urban environment provided a space in which more extreme diagrams of disciplinary power could be played out.

As such, the focus here is very much on the practical manifestation of diagrams as they were interlocked with the evolving religious communities of colonial Delhi. The CRS marked the emergence of a disciplinary diagram that could be mapped onto Delhi's urban fabric in times of communal unrest. But this, like all diagrams, was to an extent a tracing or palimpsest of prior visualisations and preparations that had accompanied the rise of communal sentiment in Delhi.

Communalism

The literature on communalism has highlighted that passionate religious feelings were not innate or primordial but can be viewed as 'instrumental' or 'constructed' (Tambiah, 1996). This does not mean that we can attribute the form and content of 'native' religions to the coloniser's hand. Indeed, Chris Bayly (1985) has shown that tensions between Hindus and Muslims had existed before many actions associated with state interference. However, religions must necessarily be affected by their conditions of existence: by the other cultures they define themselves against, by the political threats they face, by the categories they are offered incentives to identify with. Gyanendra Pandey (1990) has examined how colonial authorities focused on communalism as an inevitable division within Indian nationalism, forming part of the 'divide and rule' tactics that detracted attention from anti-colonial movements. Indian *identities* were depicted as inherently dependent upon religious affiliation and belief rather than rational and calculating thought. In India, religious communities did come to identify many people's personal and political identities, which were forged not only through the practice of faith, but also through constitutional campaigning and physical violence (Pandey, 2001).

Even if this was part of the Government's strategy, it was also well aware that excessive communal conflict in its domain would undermine not only its prestige but also its power to rule. Trade and administration suffered, but so did the image of the government as a presence that kept India in 'order' and 'civilised'. Thus, concurrent to the search for political means of control there came the essentially inseparable search for a means by which to monitor and control displays of what Peter van der Veer (1994) has referred to as 'religious nationalism'.

Narayani Gupta (1981) has provided an excellent guide to the development of relations between Hindu and Muslim groups in Delhi between the Revolt of 1857 and the capital transfer of 1911 (also see Mann, 2005a). Gupta showed how communal thought arose not only in the minds of Delhi's European population, which had moved out of the city to the Civil Lines, but also between Hindus and Muslims, who rioted during the simultaneous occurrence of their Ramlila and Bakr-Id celebrations in 1886.

Sketches of urban order: 1886–1923

The riots of 1886 caught the local police very much off-guard and had to be met with emergency and improvised disciplinary measures, including the introduction of troops into the city. These events accumulatively represented a severe *problematisation* of Delhi's urban police force, with local newspapers openly deriding their efficiency (Gupta, 1981: 133). It was acknowledged that 'special measures' would be required for festivals in the city, and these were based directly on the events of 1886.

Pickets throughout the city were referred to by alphabetical letters, as distinguished during the riots and used thereafter. This was despite their having no apparent relevance to a portion of the city or local buildings and being inconsistently used. Certain principles were also enforced, based on these experiences, over the next 20 years. First, processions (except for marriages) were banned in Chandni Chowk, the central processional avenue in Old Delhi. Second, novel processions or innovations on existing traditions were discouraged. Last, the need was established for fixed programmes for policing the city and its main routes on the occasion of the annual festivals. As such, plans were drawn up for Bakr-Id, the Moharram and Dussera (Ramlila) that 'descended to the minutest detail' and were enforced every year (Parsons, 1926: 163). While the level of detail would be overlain and surpassed by later schemes, the festival plans give an insight into the expectations of what could be known and predicted about the festivals in advance, and how the urban population could be ordered for short periods of time.

Moharram

The Moharram festival by Shia Muslims commemorated the martyrdom of the grandsons of the Holy Prophet of Islam. Various speeches and processions took place, carrying representations of the tombs of Hassan and Husein, known as *tazias*. Drinking posts and dramatic re-enactments were provided along the route, mediating the atmosphere that was both mournful and celebratory. Despite Delhi's low Shia Muslim population, the rules for 1904 noted the large number of tazias petitioned for. Processions had to be notified with a date, time and area in which they would take place. The police were given more detailed programmes, listing their pickets and dispositions for both small and large processions. Overlaying the police distributions, Magistrates were positioned throughout the city to coordinate efforts and give permission for greater use of force should the situation dictate.

The police orders attempted to ensure some degree of anticipation of violent outbreaks. Officers were encouraged to tour the jurisdictions of their barracks and report non-confidential information to the local Inspector by telephone. Thirty constables, with an extra ten men in advance, accompanied the large processions as well as three Sergeants, and two mounted *sowar* guards for communicating back to headquarters. Around 100 troops were distributed at six pickets throughout the city, while men were also placed on the roofs of houses to prevent stone throwing. The total number of men, excluding traffic police and army troops, deployed on any one day was around 150. This combination of pickets and procession attendants was hoped to cast a net of visibility over the city, linked together by mounted guards and an insecure telephone system. The technological means of force was also limited. While Hauz Kazi pickets were armed with carbines (small firearms), bayonets and 10 rounds of buckshot, other pickets had just swords and batons.

Ramlila

The *Ramlila* celebrations during *Dussera* (the 'festival of joy'), organised by the Hindu community, celebrated the life of Lord Rama through the burning of effigies and dramatic re-enactments of religious scenes. The procession was tied to a specific route that began just to the south of Chandni Chowk and then worked its way through the heart of Old Delhi to the Shah ji-ka-Talab ground outside the city walls. Ramlila was the main public Hindu festival, with *Diwali* (the 'festival of lights') and *Holi* (the 'festival of colours') being celebrated more at the community or domestic level.

   The police orders for 1904 were placed for repetition *mutatis mutandis* (with necessary changes) in future years and centred on a guard for the procession and its endpoint, and pickets throughout the city. Four pickets through the city totalling 101 men complemented the guard of 20 constables, 2 Sergeants and 1 Deputy Inspector for the procession. The troops at Hauz Kazi were armed as for the Moharram celebrations, while no special provisions were made for the other men. In total, 124 men were allotted in 1904 to police the processions, a figure that had risen to only 196 by 1922.[104] In 1927 the number of men deployed was still only 194, but these were distributed in twice as many half-strength pickets, covering a wider spread of the city.

Bakr-Id

Following the Muslim fasting month of Ramadan came the *Eid-ul-Fitr* (henceforth referred to as 'Id' in line with the colonial diagrams to be examined, unless alternative spellings were used in individual documents) festival at a specially prepared Idgah that formed the terminus of a religious and highly public procession. Cattle were slaughtered and their meat carried back to celebrations at individual homes. Since it depended on the movement of various people throughout the city, the preparations for Id were the most comprehensive of all the festivals. Two proclamations were made mapping out a revised and intensely legalistic geography for the city during the festival period, while police orders mapped out the finer landscaping of the policing apparatus. Because of the troubles in 1880 the first proclamation by the Deputy Commissioner explicitly banned the slaughter of horned cattle in the municipality of Delhi except in the slaughterhouse. In addition, cattle could not be introduced into the city or paraded apart from three tightly delimited zones, within which the use of certain bazars was strictly forbidden. Each zone had one gate in the city wall through which cattle could be brought, accompanied by no more than seven people, at which the names and addresses of the owners were noted. Once the cattle had been killed, the flesh had to be carried back through the same gates, while being covered in cloth and not being displayed in such a manner as to 'hurt the feelings of others'. The second proclamation concerned the distribution of Magistrates through every ward in the city who reported disturbances to the police, issued Criminal Procedure Codes to prevent breeches of the peace, and instructed all residents who enjoyed Government honours or emoluments to maintain order. The police force of 124 men were posted at six picket points in the city, while six mounted constables toured the city on the lookout for disturbances.

Urban sketches

While these programmes granted considerable powers to the police and magistrates, to refer to them as diagrams would be to overestimate their coherence and integrity. The programmes did not respond to Delhi's growing population; they assumed that relatively little could be known about the processions in advance and made no systematic efforts to gather intelligence on local feelings, while assumptions about what could be known were very much determined by what had gone before. These epistemological considerations also informed assumptions about the identities of the people at hand. The population was clearly dealt with in terms of communities and groups, not individuals. The riots of the 1880s were claimed to be the result of actions by 'mobs' bent on 'mischief', while the failure to quell the disturbances was in part attributed to the representative heads of the communities, whether municipal or *mohalla*, failing to control their population sub-groups.

As such, the programmes established after the 1880s represented thumbnail sketches of urban order, rather than diagrams that 'descended to the minutest detail'. Despite this, there were no major disturbances in Delhi during the 1910s, although this was as much to do with improved communal feelings and growing, trans-communal nationalist sentiment than effective policing. These sketches were, however, problematised by the clashes between Hindus and Muslims in the early 1920s.

From urban violence to the CRS: 1924–34

In the early 1920s, the Muslim League denied Congress claims that it could speak for all Indians, while the Arya Samaj asserted a return to Vedic Hindu texts and stressed the purity of the Hindu religion. Following the treatment of Turkey during the Treaty of Versailles (1919), the *Khilafat* movement sought to organise pan-Islamic protest movements across Asia, an appeal that met with particular success in north India. Hindus and Muslims were largely united in the 1919 Rowlatt Mass Movement and worked together during the Non-Cooperation protests of 1920–2.

Despite Gandhi spending the 1920s encouraging social harmony, tensions between religious communities rose dramatically in this period. This must be explained by several factors including colonial classifications, economic and cultural divisions, Congress' inherently Hindu symbolism and issues of constitutional representation. These came together in 1924 in the worst communal rioting in Delhi since the 1880s. The direct cause is still not clear, although the Government blamed Hindu nationalists for trying to convert Muslims to Hinduism.[105]

On 11 July 1924, a rumour spread that a Muslim boy had been killed and that his assailants were hiding in Katra Nil, at the north-west end of Chandni Chowk. A group of Muslims gathered in the main road and started attacking Hindu shops, while being pelted with stones from the rooftops. These disturbances spread down the Chowk until the police drove the crowds off the main roads, simultaneously attempting to stop more crowds entering the city by Sadar Bazar bridge. Army pickets and armoured cars were eventually used to disperse the crowds and they continued to tour the city through the night, in addition to extra pickets being dispersed through the municipal area. On 15 July, a further dispute broke out over which route cows should be taken to the slaughterhouse, while Hindus attacked a mosque on Egerton Road the following night. By the end of this period, 16 Hindus and 1 Muslim had been killed, and 100 and 50 people had been hospitalised of each religion, respectively.

Gandhi's fast in Delhi led to a cessation of violence across the country from 8 October 1924. Despite this, communal tensions plagued the city for the next 3 years. A clash outside the Fatehpuri Mosque in March 1925 left one dead, the funeral procession of which the following day led to further disturbances.[106] Another 3 deaths and 50 injuries followed fighting over a cow procession on 24 June 1926.[107] On 23 December of the same year, Swami Shraddhanand, a leading figure of the Arya Samaj, was assassinated. Eleven months later the man charged with his murder, Abdul Rashid, was executed, leading to rioting throughout the city.[108] The crowds overwhelmed the police at the jail and captured Rashid's corpse, touring it around the city until the authorities overwhelmed the crowds at Lahori Gate and recaptured the body. In total, three communal clashes between April 1926 and August 1927 left 4 dead and 116 injured (Hasan, 1995: 114). The Senior Superintendent of Police noted on 17 December 1927 that '[t]he people in general are so imbibed with this communal antipathy that the antagonism towards the white races which was so evident a few years ago has completely disappeared and Europeans and foreigners are generally welcomed'.[109]

Despite this, the tension still represented a major problem for the local government. Riots threw into question the ability of the administration to govern effectively, they lead to increases in crime and damage to civic infrastructure; they damaged public confidence and tax revenue; and they stretched an already taut police force to the limit. As the first section of this chapter showed the 1920s saw rapid increases of crime in Delhi Province with which the police were ill equipped to deal, especially due to the commitment of men in New Delhi.

The troubles of 1924–8 had necessitated a massive upgrading of the police precautions for the annual festivals. In the case of Ramlila, this entailed a trebling of the number of men in police pickets, from 194 in 1925 to 660 in 1927. In the same year, *Moharram* was thought to demand a police force of 514, a 357-man increase on the 1904 preparations.[110] Yet, it was the *Bakr-Id* festival that attracted the most attention, having sparked the riot in 1924 and being citywide in its scope. The proclamations remained very much the same, yet by 1927 the celebrations were policed by 514 men, an increase of 412 on the 1904 figure.[111] While fewer men were used than at Ramlila, the men were deployed more intensely and were planned in more detail than any other scheme that had yet been sketched out for Delhi. The men were placed in 55 pickets around the city, more than double the number for other festivals due to the lack of a procession route as a focus (see Figure 3.3). They were instructed to disperse trouble makers within their beat and to report any rumours back to the Kotwali.

By 1933, the Bakr-Id preparations had undergone another level of reorganisation.[112] The plans were much more clearly formulated, divided into 10 sections detailing the duties of each police subdivision within the city. Although the picket deployment had reduced from 568 to 519, this represented a similar spread of smaller pickets across the city. Their duty was specifically defined as the maintenance of peace and the collection of information from within their beats that were specified gali by gali. Plain-clothed men with bicycles were allotted to collect information and pass it back to the Kotwali. Although the strength of the striking force was increased to 166 men,

Figure 3.3   Policing for Id, 1927

over 83 in 1927, the emphasis was continually placed on forestalling their use through the collection of *information*. Station House Officers were encouraged to move through their jurisdictions to supervise their pickets while using their influence and local knowledge to quell disputes.

As against the sketches of urban discipline that had been outlined in 1904, the 1930s Id preparations had taken on the shape of a coordinating diagram. The technology was at times rudimentary; in 1927, the mounted constables were armed with cut-off polo sticks. Yet, in terms of visibility, the city had been compartmentalised into potentially transparent segments, the information from which would be passed to a coordinating centre that could rank the risk of each area and distribute force appropriately. This was to be the basis of the CRS that would force a discontinuity with the policing of festivals in the future but showed clear continuities with the evolution of Delhi's urban discipline in the past.

The Communal Riot Scheme

In response to criticisms both of police indifference towards communal tension and of actually encouraging it, the Home Secretary of India instructed the Local Governments, on 13 April 1931, that they should '... do their utmost to remove causes of friction, to take all precautions that are possible, and to spare no effort in bringing disorders under control with the least possible delay when they occur'.[113] In response, local governments revised their previous plans for dealing with widespread communal violence.

This presented the Delhi authorities with two opportunities. First, it allowed economies in the number of men being deployed in times of communal tension. As the Senior Superintendent of Police in Delhi admitted to the Deputy Commissioner in a letter of 31 July 1934, the new scheme allowed revisions in line with the actual number of policemen available in the city.[114] Second, this was an opportunity to expand the means of observing and disciplining the city. As the Senior Superintendent of Police suggested, the old plans had been drawn up when the police had 'relatively little knowledge of the city'. The Non-Cooperation movement and the Civil Disobedience campaigns had presented the Government with opportunities to extend their authority in an unchecked fashion throughout the city for short amounts of time. The ongoing threat of a communal riot, however, justified a much more comprehensive stringing together of ways of seeing and policing the city.

The 'Scheme for Police and Other Dispositions in the event of a Communal Riot in Delhi', known as the Communal Riot Scheme (CRS), was issued in July 1934.[115] The general scheme had two stages. The first stage, known as Appendix A, was prior to a riot and sought to prevent the outbreak of violence. The second stage consisting of Appendices B–D and instructions for local magistrates, was brought into play with the announcement of a riot and sought to limit its extent.

With the outbreak or imminent likelihood of a communal riot, the duty officer at the *Kotwali* (the central police station), informed 13 high-profile officers of the situation by telephone and brought the first stage into play. All City Police Stations were ordered to take up 'Riot Stations' that entailed the distribution of armed reserves and police pickets throughout the city. Armed reserves were placed at the point of exit or entry to the walled city at Sadar Bazar, protecting the Kotwali and also remaining available to be deployed from it, and waiting at Hauz Qazi for the event of an influx from, or attempt to exit from, Ajmeri Gate.

These police pickets aimed to '... localise the rioting, to guard places of worship, picket "danger spots" and prevent gang attacks in mohallas [neighbourhoods], where one community is weak, and ... to suppress rioting on a large scale and to stop crowds passing between the City and Sadar Bazar'.[116] Appendix A consisted of 22 police pickets with an average of 9 policemen per picket who collectively had to patrol their prescribed locality (see Figure 3.4). The foot constables were subordinate to the head constables who in turn answered to their Superintendents. Although generally dispersed throughout the city, there was a concentration of pickets at the west end of Chandni Chowk: the location of two important mosques (Fatehpuri and Tahawar Khan) and the pathway to Sadar Bazar across the rail lines. The Town Hall on Chandni Chowk and the Jama Masjid were also identified in Appendix A as areas in which small-scale rioting could take place.

The focus in the first stage was stated to be that of protecting the Indians from themselves, whether at places of worship or in neighbourhoods with a strong religious identity. However, looking at the confidential remarks about the small-scale patrols indicates the additional intentions of the scheme. The picket to the west of Sadar Bazar aimed to prevent an eastward flow of rioters gaining access to that area, while the three posts in Sadar Bazar were all to 'stop any influx into the city', following the struggle in 1924 to keep rioters from crossing into the city using Sadar Bazar bridge. The area around Fatehpuri Mosque was not only one of communal tension but it was also the area in which enraged rioters could gain access to the centre of the city, which is what the pickets at Hauz Qazi aimed to prevent on the route from Lahori Gate. However, the post at Dufferin Bridge towards the north aimed to prevent any movement from the City into the Civil Lines, which housed the Delhi Administration and the local elite population. These inscriptions denote a clear hierarchy of priorities in terms of urban discipline; the aim was to keep rioters out of the highly populated and politically volatile old city, but not at the cost of threatening the Civil Lines to the north. Even without the boundary walls of a total institution, the scheme incorporated elements of 'enclosure' to protect what it defined as the most valuable sector of society.

Were the situation to worsen, the second stage was to be brought into play. The Army would be contacted in the Fort and ordered to move to a rendezvous point in the city in line with the Local Alarm Scheme that had been revised in the early 1930s. Appendix B stipulated that 24 pickets with foot patrols were to be established to break up and arrest rioters. Although employing a similar technique of discipline, the picket with a short 'beat', Appendix B (see Figure 3.5) marked a shift in focus from areas in which violence could be *prevented*, to those which should be *protected*. While the first stage mainly targeted those areas likely to spark off communal violence, stage two had to control a full-scale riot. Three extra pickets between Kashmere Gate and Queen's Gardens protected the Civil Lines while three pickets in Faiz Bazar and four pickets stretching from Ajmeri Gate to Paharganj countered movements to New Delhi. Beyond protecting the European enclaves, the pickets also targeted vulnerable or volatile sites throughout the city. Strong pickets protected the economically valuable

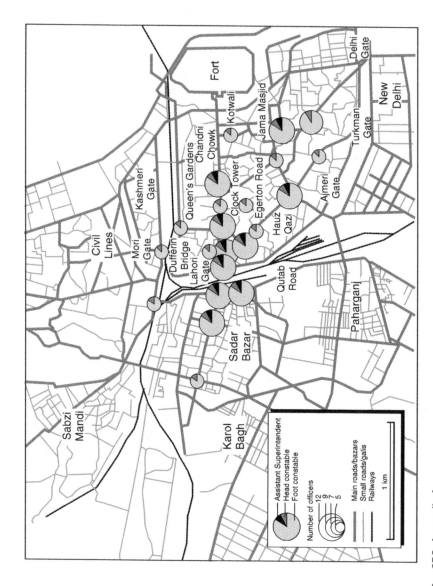

Figure 3.4 CRS, Appendix A

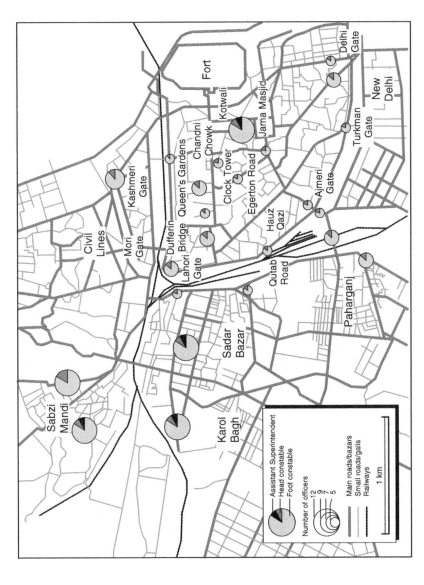

Figure 3.5 CRS, Appendix B

industrial area of Sabzi Mandi, not only addressing problems at the mills but also stopping mill hands from passing into the city. A picket to the north aimed to control crowds bringing corpses to the mortuary, following the violence after the execution of Abdul Rashid in 1927. Pickets also targeted the butcher community, who had been aggressors on behalf of the Muslim community in the past, the politically mobile student populations of Hindu and St Stephen's Colleges, and nearby jewellers and arms dealers.

These isolated pickets were united by the motor patrols of Appendix C. While the Id preparations of 1933 had distributed many pickets through the city, there was no systematic way of collecting their information without deserting their posts, which was forbidden. The CRS established a series of networks that would ideally ensure that information could flow up from the patrols to the Kotwali. Nine motor patrols were established that linked between two and three of the areas patrolled by police pickets, each being armed with muskets for more serious trouble (see Figure 3.6, in which each separately marked route marks a separate motor patrol). These three levels of control (appendices A–C) were overlain by three motor patrols of local magistrates; one for south Old Delhi, one for north-west Old Delhi and one for Sadar Bazar (see Figure 3.7). These covered much larger areas than Appendix C patrols and were designed to pass information to and from the Kotwali.

As the developments between 1924 and 1934 had shown, the visibility of the city depended upon the technologies available. The CRS made full use of motor patrols to ensure the swift movement of information around the city, while the phone network was also used to alert the city and the Fort immediately if riot status were taken up. By these means the city was, ideally, segmented into knowable and thus controllable parts, all unified by a system of surveillance.

However, beyond the specificities of spatial technologies were the juridical, more general techniques of control. Appendix D stated that as soon as stage two was enforced, the Senior Magistrate at the Kotwali would be granted a whole arsenal of orders and proclamations. 'Subsection a' was described as advisable and focused on establishing the Kotwali as a centre of local knowledge while 'subsection b' looked at ways to consider bringing the wider spaces of Delhi under control. Under 'subsection a' the Kotwali was made the best-equipped centre for coordinating the information that the surveillance system provided. This was achieved through collecting together certain advisors, including Stipendiary and Honorary magistrates, *tahsildars* and *chaprasis* (local officials), the Health Officer, Municipal Officer, Civil Surgeon, Assistant Engineer in charge of Telephones, Superintendent of the Fire Brigade and town criers. While these men would offer practical advice, the flow of information into and out of the Kotwali was managed by the CID staff, all of whom had to report for duty. Of these, 24 men were

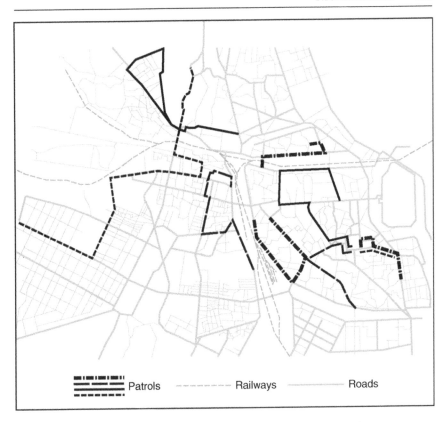

Figure 3.6   CRS, Appendix C

directed to move about the city, collecting information and passing it back to the Kotwali.

'Subsection b' also stipulated the means by which the Kotwali could extend its influence throughout the city without necessary falling back on the presence of the police. These included more personal measures, such as establishing conciliation committees, requesting locally respected gentleman to accompany police pickets, and deploying other influential men in times of need. Wider reaching measures aimed to stop the situation becoming inflamed, including the 'complete control of the press' by vetting of all articles written, the picketing and prescription of volatile funeral processions, the use of a curfew or of Section 144 of the Criminal Procedure Code banning the carrying of weapons or assemblages of more than five people. More aggressive measures included persecuting those engaging in exciting ill feeling, or even those suspected of being 'bad characters' and authorising Magistrates to give the order to fire in self-defence.

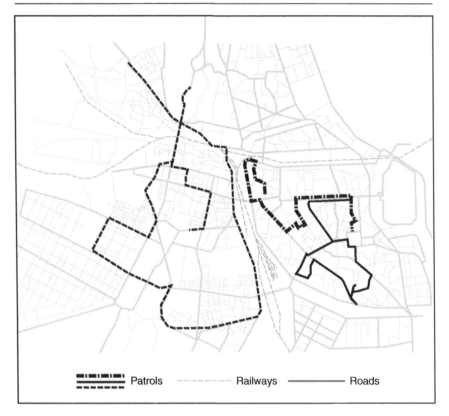

Figure 3.7   CRS, Magistrate's Patrols

Thus, in theory, the local administration had the ability to enclose parts of the city in accordance with a disciplinary diagram in which space was segmented into controlled and knowable parts, in which power was ramified and knowledge collected by touring patrols, under the coordination of the all-seeing eye at the Kotwali. This marked an epistemological shift in terms of belief in what could be known about the urban sphere. The CRS was the diagrammatic embodiment of governmental faith in the means of urban discipline to penetrate urban populations and make them, if not docile, then at least obedient. Past experience was used to inform the distribution of troops that mapped out the geography of risk across the city. The analysis of identity remained at the level of the community leader and was steeped in terms that relayed the mob mentality of the crowd.

The biggest shift was in the realm of visualising the city. As the map attached to the back of the CRS booklet highlights, the city was cartographically displayed in a police programme for the first time (see

Figure 3.8). Here, the city was shorn of major topographical features and displayed as a site for the distribution of force. This geography of the old city was compounded by laws that took in the whole area, aiming to control speech, movement and grief. The total number of policemen and magistrates distributed through the city was 341. While this was over 300 people fewer than the Ramlila policing of 1927, the organisation sought to make the system more effective. This depended upon developments within the colonial techne. These involved not just material instruments like the telephone or the motor car, but the art of using these technologies, orchestrating them so as to channel the flow of information around the city, predicting and diffusing violence, ideally, before it occurred. All this within a milieu suffused with the ethos that the European enclaves, and the capital especially, be protected at all costs.

Following the issuing of the CRS, the number of men used to police the annual festivals dropped, while the policing programmes started to take on the forms of the disciplinary diagram outlined in the scheme. Analysing the annual festivals helps to highlight not only the diffuse impact and workings out of urban diagrams, but also how their panoptic urges were often further problematised and resisted when put into practice.

Diagrams through festivals: 1936–46

Following the issuing of the CRS in 1935, the number of policeman on pickets and patrols during religious festivals fell from averages of between 450 and 650 per year to about 250. This was due not only to increased efficiency, but also to the fact that the Senior Superintendent of Police knew that in the case of a riot he could call the CRS into effect. Despite this, the festival preparations continued to develop in their sophistication, as well as having to adapt themselves to a Sikh tradition that became radicalised in the late 1930s.

The effect of the CRS on the festival preparations was as noticeable on the page as it was on the street. The 'Orders and Dispositions' that were created annually to tailor the schemes, *mutatis mutandis*, were systematised. These primary spatialisations had previously been a collection of letters, typed picket lists and notes on procedure, but from 1935 took up the appendices format of the CRS.

The *Moharram* preparations for 1935 saw a reduction in police allocation from around 510 during an average day of processions in 1929, to 380.[117] In line with the CRS emphasis on mobility and the flow of information, more police accompanied the procession and passed information through their officers to the magistrates. They were instructed by the Senior Superintendent of Police to '... make due liberal allowances for minor faults

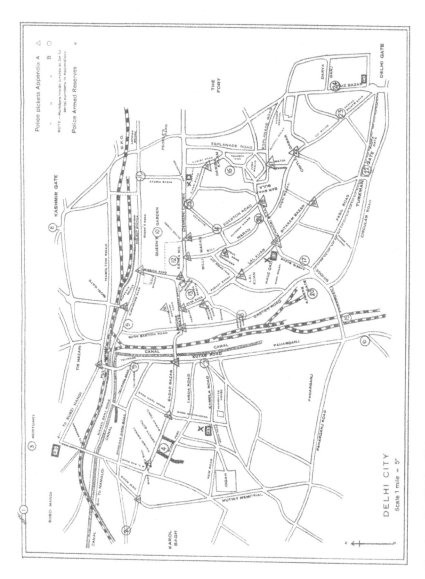

Figure 3.8 Map from the CRS booklet

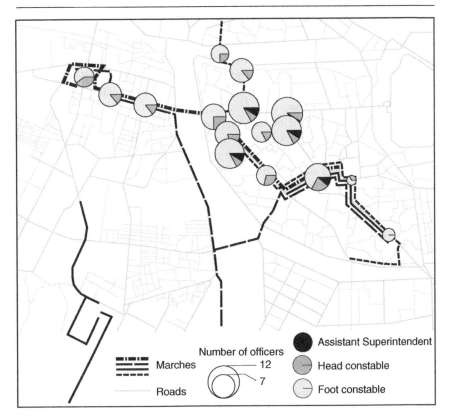

Figure 3.9   Moharram processions and policing, 1935

committed by the people engaged in performing the ceremonies attaching to the festival'. In 1943, these instructions were augmented by patrols that would collect extra information from throughout the city and unify the pickets, although these were only playing a minor role by this point (see Figure 3.9).[118]

In 1935, the first Moharram procession to pass through New Delhi had taken place, being celebrated by the Bengali clerks that had come to Delhi from Calcutta after 1911.[119] The processions was sanctioned and allotted a Magistrate to supervise the celebration. The introduction of another procession in the capital in 1941 caused the Chief Commissioner some angst. He confided to the Deputy Commissioner that he would prefer to stick to the 'no innovations' policy as formulated after the riots of the 1880s, but that exceptions must be made for a newly constructed area.

The plans for *Ramlila* became increasingly similar to those of Moharram from the late 1930s onwards. The emphasis in 1935 moved away from pickets and towards patrols through the city and an escort with the procession.[120] Similar orders were issued to the men, the number of whom dropped from 660 in 1927 to 390 in 1934 and 247 in 1943. By this time the Ramlila celebrations had introduced a procession into the clerks' quarters of New Delhi, for which Traffic Officers were put on special duty to police the border between the two cites.[121]

The preparations for Id obviously resembled the CRS it had helped to inspire. In 1935, the number of allotted policemen dropped to 401, from 660 two years earlier. The number of men remained at this relatively high level into the 1940s due to the need to police the whole city. The same proclamations and zoning were used as in 1904, but with a wider spread of pickets and the use of intense traffic policing around the slaughterhouse and Idgah in Sadar Bazar to channel the processions.[122] There were over 70 more foot constables employed on Appendix A pickets than in the CRS and more high-ranking officials organising them. The pickets have more of an emphasis on Bazar Chitli Kabar and north of the railway lines (see Figure 3.10), in order to compensate for the increased activity in these areas in which cattle could be stored.

As such, the policing of the traditional festivals took on the features of the CRS in terms of administration, a reduction of men and an increased emphasis on mobility of patrolling the city. However, it was with regard to the new policing of an annual ceremony that the most sophisticated disciplinary diagram was formulated. The Sikh community in Delhi commemorated the martyrdom of Guru Tegh Bahadur who, in 1675, was killed by Emperor Aurangzeb. According to local tradition, the Gurdwara Sisganj marks the spot in Chandni Chowk at which Guru Tegh Bahadur was beheaded for his religious dissidence. An annual procession transferred the *Granth Sahib* (a religious text) from the Gurdwara Sisganj on Chandni Chowk to a different Gurdwara, either Rakabganj or Bangla Sahib in New Delhi.

The commemoration had undergone a transformation in the 1930s from an occasion for Sikh celebration to anti-Muslim demonstration. This took the form of gathering outside mosques and creating disturbances during the daily prayer times. In 1937, the Sikh leaders promised not to enter Paharganj Bazar until 8 PM but did so and played music before the end of the Isha prayers at the nearby Mosques, as happened again in 1938.[123] In 1939, the police refused to accept verbal assurances and insisted on a clause being written into the Police licence for the procession that music should not be played before the end of mosque prayer time. The authority of the police was again defied such that, as the Deputy Commissioner stated on 30 November 1940, the city was 'brought to the brink of a communal riot'.[124] To avert this in 1940, a complex series of policing orders were constructed regarding the commemoration.

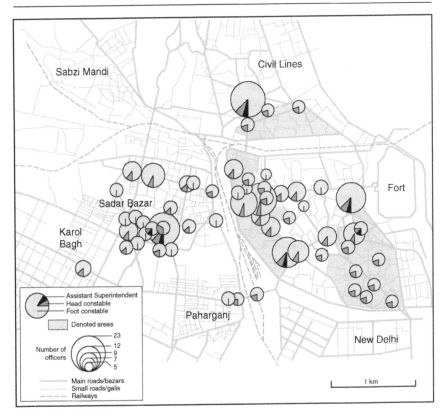

Figure 3.10   Id pickets, 1940

The plans were extensive, including troops, pickets and police escorts. Army troops were positioned at the exit from the city and the heart of Paharganj Bazar along the route, while the Jama Masjid, New Delhi Police Station and New Delhi Railway station were afforded protection regardless. Special note was made that the procession would terminate at the Gurdwara Rikabganj in the heart of New Delhi, near the Viceroy's House. Special army troops were positioned to prevent any 'unruly elements' approaching the Secretariats or Viceroy's residence. Each of the four main Police Stations in Old Delhi distributed a total of 129 policemen within their wards. This distribution was based directly on the CRS, using the 'serial numbers' to denote locations in the city from Appendix A.

As such, the traditional use of army troops was overlain with the more disciplinarian ordering of space outlined in the CRS, in which pickets were placed throughout the city in the hope of making it transparent and know-able. Yet, this system itself was further developed through adapting it to the local environment. Special pickets were created for the 14 mosques along the route, formed by 23 head constables and 144 foot constables, as depicted in Figure 3.11.

The duty of these pickets was to ensure there was no stone throwing or violent assaults, while it was explicitly stated that 'Offenders who, after due warning, persist in making murderous assaults and if they cannot be dealt with otherwise or sufficiently promptly may, under the orders of the Head Constable, be shot at.'[125] Mounted police patrols also cleared the roads from 3 to 6 PM while an armed police force escorted the procession behind the gas squad lorry that proceeded well ahead.

These preparations highlight the fact that Delhi did not just have a religious geography, but also a religious time–space geometry. The Muslim prayer times of *Asar* (defined in 1940 as 4–4.30 PM), *Meghrib* (5.45–6 PM) and *Isha* (7.30–8.30 PM) all coincided with the procession's duration. This was not just a geography of religion, but also one of sound and culture. Prayer required silence and contemplation, whereas 'festival' warranted noise, music and celebration. Even if the mosques were not physically violated, their sacred space could be impinged upon from without. The authorities had tried to seek assurances that the processions would not enter certain areas until certain times but had met with little success. Due to this, in 1940 detailed timings for the procession were issued to ensure that it would not reach the sensitive mosques on route at times of prayer.[126]

Were the processionists to pass along the route at a steady pace they would, as the Deputy Commissioner noted on 30 November 1940, have passed through three sacred spaces (collections of mosques in Egerton Road, Ajmeri Bazar and Paharganj) at three sacred (prayer) times.[127] This time route is expressed by the solid line in Figure 3.12. As such, the licence extended from 4 to 11.30 PM but with explicitly stated intermediate timings, as represented by the dashed line. The procession was not to enter Egerton Road before 4 PM, thus avoiding the mosques there during the prayer time of Asar. It also had to pass through Ajmeri Gate Bazar between 6 and 7 PM, avoiding Meghrib prayer time, and was not to enter Paharganj before 8.30 PM, thus missing the Isha prayers. Ideally, the procession would move through the city at a regulated pace, avoiding those places with clusters of mosques at times of prayer (see the 'licenced rate' in figure 3.12). As such, the disciplinary arts of distribution reformulated in the CRS became re-articulated in the diagram of urban control that could bring spatial and temporal order to the annual procession.

Resistance and transgression

Thus there is no diagram that does not also include, besides the points which it connects up, certain relatively free or unbound points, points of creativity, change and resistance, and it is perhaps with these that we ought to begin in order to understand the whole picture. (Deleuze, 1988: 44)

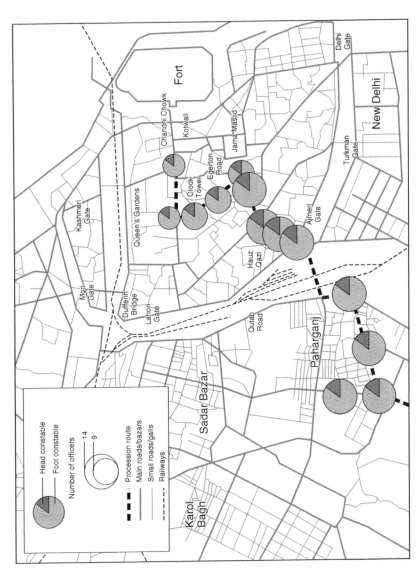

Figure 3.11　Pickets along the Guru Tegh Bahadur Commemoration Route, 1940

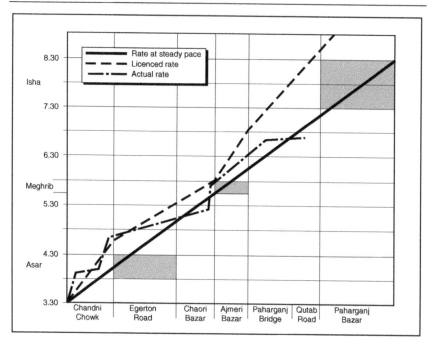

Figure 3.12   Time–space Routes for the Guru Tegh Bahadur Commemoration Procession, 1940

The 'messy actuality' involved in disciplining Delhi's festivals provides ample historical support for O'Malley et al.'s (1997: 509) assertion that we should move away from studying the mentalities of rule and focus on the implementation of diagrams and technologies. The partition of the urban landscape into controllable segments was contested in the 'orgiastic' festival atmosphere during which, Veena Das (1990: 19) has suggested, 'there is the potential of interrogating and mocking the normal social arrangements of power and hierarchy'. The movement of people around the urban form during these festivals gave bodily expression to Foucault's (1980) assertion that power cannot be possessed or canalised, rather it is mobile, transitory and open to 'creativity, change and resistance'. The following implementations of the disciplinary diagrams not only illustrate this resistance, but also show how it did become internalised through programmes of review that updated and modified the diagrams themselves in the increasingly acrimonious atmosphere of the late 1930s.

On 30 June 1937, the music from a marriage procession had offended those offering prayer at Farrashkhana Mosque, leading to a street fight in which 24 Hindus were injured. Similarly, on 1 July 1938, band members in a Hindu procession refused to acquiesce to Muslim requests to stop playing music outside a mosque, leading to another violent attack.[128] On 26 August 1940, the Paharganj Police Station House Officer, a Sikh, was stabbed to death while trying to control a Hindu procession.[129] These outbursts of violence were portentous of the troubles that would follow in December 1940 with the grander scale annual processions.

The scale of the protest tended to match the scale of the festival, and displayed how deeply intertwined the ceremonies and policing had become. During the *Moharram* festival of April 1938, the Senior Superintendent of Police decided not to form a cordon with rope around the mourners, as had been the previous policy.[130] Rather than celebrating this retraction of police control, the processionists resented the move and squatted on the ground near Chitli Bazar, in the centre of Old Delhi. The police eventually agreed to escort the Tazia, but by this point they were behind schedule and clashed with another procession. Fighting broke out between the groups and a Magistrate declared the assembly unlawful. The crowd ignored the proclamation and had to be dispersed by force. A much more complex interweaving of anti-colonial and communal acts of resistance took place 2 years later during the implementation of the new plans for the commemoration of the martyrdom of Guru Tegh Bahadur. This serves as a sharp reminder that the disciplinarian ambitions of the colonial government were not always transformed into reality. Rather, the attempts to create routes and barriers to channel the procession produced as many spaces for dissent as for control.

The licence for the procession of 4 December 1940, it was later discovered, had been taken out in the name of a 'poor person of poor status' who was not on the Gurdwara Committee, thus allowing the Police order to be violated.[131] The processionists attempted to start an hour before the agreed time of 3.30 PM and were cordoned in by the police. Although eventually starting on time, the attempt to keep the flow and noise of the procession away from the Muslim places and times of worship was a shambles. A.P. Hume was called on duty that night and recorded the progress of events in a letter to his parents.[132] He recalled that the Deputy Commissioner had great trouble 'shepherding' the 10,000–15,000 Sikhs through the city along what he called 'a strategic line of defence'. As represented diagrammatically as the 'actual rate' in Figure 3.12, the Deputy Superintendent of Police and the City Magistrate jointly reported on 11 January 1941 that

> [t]he processionists who under these circumstances, started at about 3.30 and were almost stationary in Chandni Chowk for half an hour rushed and tried to reach in front of the mosque in Nai Sarak [Egerton Road] at evening prayer time. They failed to do so. Next attempt of the processionists was to reach the mosque in front of the Gali Shah Tara at the next prayer time and the distance of about 2 miles was covered at such brisk pace that the procession reached

Hauz Qazi Chouk before time mentioned in the license. Danger was that there would be trouble in front of the mosque. Dilatory tactics were adopted. Much of criticism and even threat that the processionists would squat, was faced, but they were not allowed to go beyond the Chouk till after prayers which had been arranged to be said a few minutes before time.[133]

A nationalist political worker attempted to stir the crowd outside Ajmeri Gate, claiming that the police escort for the Granth Sahib was a disgrace. The crowd, thus roused, endeavoured to overtake the Police three times in an attempt to reach Paharganj before the final prayer time of 7.30–8.30 PM. Hume, who had been positioned nearby, joined the crowd to assess the situation. At the front of the 'solid mass of humanity', he found a double rank of police with lathis facing the head of the procession, 'whose main job was to time the movement of the processionists so that no Sikh band come opposite a mosque at prayer time'. However, the procession spilled over Paharganj Bridge and there were not enough police to check the various routes through which the crowd emerged. Despite the polices' efforts, the crowd went on surging towards Paharganj. It was only with the arrival of the formidable tear gas squad that the crowd abated and were kept away from Paharganj Mosque, although they later went on to jeer at the Muslims in Paharganj Bazar.

In the future neither the Muslim prayer times nor the Sikh procession could be sufficiently altered, so after a programmatic review, the only alternative for the local administration was to make its forces as mobile as the processionists. The review of the 1940 procession concluded with advice to prevent the recurrence of such events. The licence should be in the name of a respectable man, it suggested, the time of the procession should be strictly adhered to, certain 'fire-brand *Akalis*' (martial caste Sikhs) should be banned and more police should be allotted to the procession. All of these suggestions were built upon in the orders for the procession in 1941. A licence was issued under Section 30 of Act V of the Police Act (1861) to the President of the Gurdwara Parbandhak Committee, making them officially responsible for any transgressions of the licence provisions.[134]

The actual pickets at the mosques were not strengthened, although extra men were placed on rooftops near mosques along the route, while military posts and mobile police reserves were positioned throughout the city. However, in an attempt to keep the procession at a set pace, the escort was reorganised and strengthened to ensure a strictly disciplined, mobile operation. Magistrates and 120 policemen accompanied the procession, with technological support from the tear gas squad, a concertina wire squad with magazine rifles, mounted police and motor lorries carrying armed reserves.[135] Other troops to martial the procession brought the escort up to a total of 191 men. Were a riot to break out the instructions were to use as little force as necessary, to precede the release of tear gas with a bugle or loud speaker, to likewise warn against the use of controlled fire, of which the exact details of the number of rounds fired, the time, circumstances and names and addresses of casualties and actions taken for their 'succour' should be recorded.

In 1941 the Gurdwara Committee refused to accept these further restrictions so the licence was withdrawn and CrPC 144 declared at the time of the procession. Over 2,000 people gathered in the south-east of Queen's Gardens where the Gurdwara had constructed a ceremonial *pandal* (meeting ground). Police lorries were sent down Chandni Chowk at 11 AM warning the crowds to disperse, while a Magistrate's order and verbal persuasion were employed at 1 PM.[136] The failure of these measures and the alleged attacks made on mounted police convinced the authorities of the need to use tear gas. At Queen's Road, Queen's Gardens and Fountain Square, 30 rounds were released while, in response to a 'procession' emerging from the Gurdwara Sisganj, over 70 rounds were released, completely obscuring the street.[137] The total assault unleashed 144 shells and 15 grenades of tear gas, a ricocheting canister of which killed a child in the crowd.

In the government report on the firings, sufficient warning was said to have been given and approval was granted for this experiment with new time–space ordering and the new disciplinary apparatus of the tear gas squad. As stated in the Chief Commissioner's Fortnightly Report, 'The complete collapse of Sikh truculence, which had at one time attained a dangerous pitch of fanaticism, and the satisfactory outcome, is to be ascribed very largely to the employment of tear smoke.'[138] Some members of the police also shared these feelings. J.F. McLintic later recalled:

> It was a memorable experience, on this day when tear gas was first used, to find that when the smoke had drifted away one could walk along a silent, deserted street which a minute earlier had been packed by an angry, howling mob. The debris of flight was the only evidence of their passing, the litter of shoes and garments bore witness to their haste. They made no attempt to reassemble; they retired from the fray, not to lick their wounds, but to dry their streaming eyes.[139]

Such force had not been used in retaliation to a great deal of violence, although the authorities did report stone throwing and the jostling of the police members trying to disperse the crowd. Instead, the measures were deployed to enforce a pre-defined ordering of space, and time, that had dictated the disciplined procession would only take place on its terms. As the *Hindustan Times* reported on 24 November 1941, unintentionally recalling a militaristic forefather of the arts of discipline: 'The Gandhi Grounds, where a Diwan of the Sikhs was held, presented the scene of a besieged fortress. It had been cordoned off from three sides, and isolated from the rest of the city since early morning. ... The whole of Chandni Chowk from the Red Fort right up to Ballimaran presented the appearance of an armed camp.'

The policing scheme for *Id* remained the same throughout the 1940s, with the local authorities seemingly able to maintain a balance between the restrictions on cattle parading while allowing the tradition to survive in the old city. However, trouble arose when such traditions were introduced into new urban areas where there was no precedent to decide which religious community's sensitivities were to be catered for. On 6 November 1946, a fracas erupted over a cattle procession for the Id festival through a recently built-up area of Paharganj for which there was as yet no prescribed route.[140]

The unfolding events saw some of the non-Muslim population of the district extend their territorial claims from the residential to the public realm through contesting the movement of the procession from the Mosques of Paharganj to the Idgah of Sadar Bazar. Police had been allotted for the Paharganj region while the procession itself was accompanied by armed patrols, mounted police and a tear gas section patrol. In attempting to proceed along the traditionally used Chuna Mandi Road, various Hindus and Sikhs, including the high-profile Congress politician Desh Bandhu Gupta, attempted to persuade the processionists to use a new route up the more-recently built Chitra Gupta Road. An angry crowd of 200 Hindu's gathered, after which the procession adopted the new route, which was then barred by a group of Hindus and Sikhs (there are no reports on how provocative the Muslims were being at this point). Abuse, shouting and stone throwing ensued, so W.D. Robinson, the Senior Superintendent of Police, decided to use tear gas. However, the dense network of lanes and galis rendered the gas ineffective, while people also moved up onto the rooftops. As the stone throwing continued, even from the nearby government quarters, Robinson gave the order to open fire. A curfew was put in place, but the communal tension in the city continued to be unleashed for over a week, including serious clashes in Sadar Bazar on Wednesday, 13 November in which 10 were killed and riotous mobs spread to Daryaganj, Faiz Bazar and Kashmere Gate. By 20 November, 28 had died and 54 were seriously injured.[141]

This had not just been a local disturbance, or one that spread through a city, it had taken place in the capital of India, with Paharganj technically being part of the New Delhi Municipality. The Home Member of the Government of India himself criticised the local authorities in a note on 17 November for the way they handled the affair, questioning why the magistrate had not set out a path beforehand when he knew the area was one of communal tension, why 200 Hindus were allowed to gather, and then how a cow was introduced to spark off a highly sensitive situation?[142] High-ranking officials were brought in too late and even after the event precautions were not sufficient to prevent a wave of stabbings throughout Sadar Bazar. Despite the reports that stones had been thrown from Government of India quarters, it was also mentioned that '[t]here are allegations that persons living in Government quarters were not properly treated ...'.[143]

This tart reminder that the Delhi police were protecting the heart of Empire, not simply a local district, prompted a review typical of what Rose (1996: 42) referred to as a 'problematics of rule' during which resistance is internalised in an attempt to anticipate its further recurrence. Suggestions for dealing with such situations in the future were made that guaranteed the tighter politicising of religious festivals in the period of communal tension that preceded Independence and Partition.

The suggestions can be divided into the following four aims with their correlative techniques. To placate the city in times of tension, peace committees would be formed in each mohalla, broadcasts would be made on communal harmony and the mutual celebration of each others' festivals would be encouraged. In terms of intelligence, suspects' houses would be searched, 'bad characters' would be listed and militant communal organisations would be placed under observation. Enforcing strictly patrolled curfews, deploying the Criminal Procedure Codes, intensely patrolling strongly communal mohallas, and placing watches on mosques and temples to detect 'doubtful strangers', would increase surveillance of the city. Finally, prosecution was encouraged in cases of violent assault or of newspaper hawkers spreading inflammatory rumours. Such powers sought to render the capital secure, the need for which was reiterated by a District Magistrate on 20 November 1946:

> We must not forget that the administration of Delhi has a heavier responsibility for Delhi is the seat of Government. Anything which happens here has repercussions on the rest of India. Any disturbance here attracts more attention, not only in this country but in the whole of the world. Moreover, if work in the offices of the Government suffers, as it has suffered due to disturbed conditions during the last fortnight, the entire economy of the country will suffer.[144]

The CRS was the most comprehensive attempt to shoulder this heavy responsibility, achieved fundamentally through an epistemologically reconception of Delhi's urban places as political space. The distribution of stationary and mobile forces throughout the city sought to make the dark and winding galis of Old Delhi visible, creating spaces of the law and of violence supported by the technologies of discipline and surveillance. This not only drew on the local administration's imaginary geographies of risk and the practical development of its technologies of discipline, but was also linked to the 'boundary and Keep' ethos of New Delhi. While one disciplinary system spanned the two cities, another worked to define and police the boundaries.

These programmes marked a regime of government that spanned the two cities, attempting to secure the capital not just as the home of the sovereign power, but also as a disciplined and visible space. The nature of this government has been shown to be in turns fragile and violently powerful, possessing both penetrating vision, yet also being blind to the cultural sensitivities and winding galis of Old Delhi. The imposition of order was resisted at all stages, whether through mocking commentary, petitions over inefficiency, criminal activity or full-scale riots. Rather than operating silently and creating docile subjects of political obedience, the systems of urban discipline in Delhi repeatedly had to resort to sovereign mechanisms of violence in times of emergency. These actions possessed their own logic that remained stubbornly external to the government, while remaining fundamentally constituted by, and constitutive of, the colonial administration.

As such, the procedures outlined here present a microcosm of the disciplinary practices of colonialism in India. Operating without the liberal checks of Europe, the use of violence was justified in terms of the necessary restoration of order amongst a population that only understood the language of force. As such, the colonial policing and military systems represented the indissociable combination of disciplinary surveillance and ordering, sovereign violence and laws, and governmental regulation and conduct of conduct. These articulations advanced during times of crisis and problematisation brought about by breakdowns in public order. Such breakdowns also occurred in the biopolitical sphere, understood as the rise of 'congestion'. This led to advancements in the technology of urban improvement that sought to regulate the processes of population and urban growth, as detailed in the following chapter.

Chapter Four

Biopolitics and the Urban Environment

The growth in Delhi's population had reached such a level by the 1930s that it was threatening the security of the capital region epidemiologically and in terms of spatial order. This chapter examines the ways in which the government attempted to regulate the population of Delhi city through urban improvement. This imbricated an interest in the health of the population, through the biopolitical domain of the governmental pole of biopower, with sovereign power over land rights and territorial constitution. This combination sought not only to make the population productive and healthy but also to quell the growing political criticism of such an obviously malfunctioning urban space.

The inability of the government to sufficiently *visualise* its population from the 1860s to the 1930s, and the effect this had on the health of the city, will be examined as a prelude to three investigations of the ways congestion *problematised* Delhi's urban governance (see Table 1.1). The second section will examine the formation of the Delhi Improvement Trust (DIT) as the *techne* through which congestion would be tackled, incorporating *identity* assumptions about the slum dwellers and the way their needs were to be calculated. The third and fourth sections will examine the two largest projects of the DIT, both of which failed to certain degrees, as a means of exploring the ways in which the Government sought to avoid the financial consequences of securing the health of its dependent population. The arguments will be framed through an analysis of biopolitics as one of the domains by which governments sought knowledge of, and control over, their populations.

Population Expansion and Urban Disorder

Domains of government

As the introductory chapter stated, eighteenth-century European governmental thought and practice conceived of domains that were necessarily external to the state, including the 'economy', 'population' and 'society'. Despite their supposed autonomy, fears about the stability of these realms led to intensive investigation to detect the conditions that best guaranteed their functioning. Parallel to the emergence of political–economic beliefs in self-regulating economic forces, and the belief in a necessarily independent 'society', biopolitics was studied through particular, external 'domains'. It studied the demographic patterns of births and fertility, the nature of endemic diseases, the prevalence of biological disability and the effects of the environment on human life (Foucault, 1975–6 [2003], 243–5). Though only detectable at the abstract level, the processes in these domains could be targeted through interventions in the health and lives of people on the ground. For instance, economic stability relied upon people making informed and well-calculated investments, stable social processes relied upon conscientious, self-regulating citizens, while biopolitical stability required rational calculations about accommodation, reproduction and diet.

European biopolitical regulations were radically transfigured when applied in colonial contexts (see Legg, 2006a). Yet the need for a healthy, and thus productive and content population was still central, although the subjects of this population were deemed to be irreducibly different. This latter assumption led to a lesser reliance on the liberal governmental 'distanced' approach to biopolitics that encouraged conduct, and more of an interventionist approach, treating individuals like resources to be managed.

Gyan Prakash (1999: 127) has stated with regard to India that '[t]o govern Indians as modern subjects required colonial knowledge and colonial regulation to function as self-knowledge and self-regulation, but this was impossible under colonialism'. The attempt to bridge this gap took place in spaces of sanitary regulation, epidemic controls, statistical management and organisation of the population that Prakash (1999: 144) terms, following Partha Chatterjee, 'political society'. This zone of mediation between the state and individual was both wider encompassing and less intense or elitist than civil society. It was empirical and descriptive, based upon the concept of population, and allowed the government to reach the inhabitants of a country as the targets of 'policy' (Chatterjee, 2000: 24, 2004). As India became increasingly urbanised in the twentieth century, this biopolitical policy became one that was ever more concerned with the 'urban question'. Thus, while the colonial government was biopolitical, it did not extend the full range of liberal governmental tactics into the alien Indian population.

This was not solely due to a callous indifference to native life. Rather, colonial biopolitical aspirations had to contend with colonial economics that demanded a profitable state. The temporally distant financial remuneration from investments in the population's health often led to conflicts between rationalities from different governmental domains. The colonial world has long been acknowledged as one of resistance, ambivalence and rupture. Orientalist representational structures and colonial administrative practices were resisted and criticised, while these resistances themselves were ruptured with internal divides and continuities with the worlds they sought to reject (Nandy, 1983; Chatterjee, 1986). Edward Said's suggestion of a stable colonial self constructing the oriental other has been revised, highlighting the fraught and unstable nature of the colonialists themselves (Bhabha, 1994). Beyond these individual and psychoanalytical strains were the broader tensions of empire that accompanied the day-to-day administration of the colonial world (Stoler and Cooper, 1997).

The regulation of simultaneous governmental domains made demands on different types of resources and when these demands clashed, certain problematisations of government would arise. These problematisations were raised, negotiated and defended by certain people. Again, these tensions did not have to arise between governments and an external party, they were often internal and, indeed, were the key to keeping programmes of government up to date and rigorously designed.

One of the most common clashes was between the economic stringency of the administration and the demands made by the increasingly complex and dangerous industrial world. The urban sphere was a cauldron of competing claims for regulation and intervention, a sphere that governments both feared and desired for its financial and political potential, and costs. As the congested capital of British India, Delhi needed intervention, but the degree to which the government was willing to intervene would depend upon the tenacity and energy with which the demands for biopolitical regulation were put forward against the governments competing demands for financial stringency. The clash between finance and health in Delhi was a recurrent one and surfaced at the denouement of some of the DIT's biggest projects in the late 1930s. The 'underfunded and overextended' (Prakash, 1999: 13) nature of Delhi's biopolitical apparatus had, however, long been apparent.

Planning without vision

Jyoti Hosagrahar (2005) has examined the everyday processes by which urban planning was resisted in Delhi between 1857 and 1910. Between the end of the 'Mutiny' (1857) and 1887, almost one-third of Delhi's urban landscape was destroyed. This was the result of the Government of India assuming the sovereign powers and possessions of the Mughal Emperor, the local territory of which was referred to as the *Nazul* lands. In this time the Delhi Municipal Committee (DMC), established in 1864, also worked on civic monuments, urban infrastructure and attempted to establish regulations throughout the rest of the city. At this level, the government intruded upon what were previously local decisions about public space and met resistance in various forms. The weak and disempowered made room for themselves in quiet, everyday actions, public services were vandalised, buildings were erected without permission and defended in court, and public space was illegally rented out (Hosagrahar, 2005: 71). In terms of more organised opposition, the influential jewellers of Dariba had effectively argued against the destruction of their community after 1857, while by the 1860s, petitions were made against the DMC's failure to deal with open drains and epidemics.

Vijay Prashad (2001: 125) has charted the continuation of these protests in relation to the technology of sanitation in Delhi. In April 1871, the Urdu Akbar newspaper criticised the DMC for failing to provide clean canals, gutters or drinking water in the city. Plans to destroy the city wall near Lahore Gate were resisted in 1881, and the Municipal Committee was criticised again in 1894 for failing to sufficiently administer such a 'famous city' while the Europeans in the Civil Lines were well catered for. The tone of these complaints would continue into the next century, but their nature was completely transformed after the Durbar announcement of 1911. While most of the attention focused on the new city, the old one was not ignored in the initial rhetoric and planning. Viceroy Hardinge himself outlined the aesthetic burden that Old Delhi would bear in his address to the DMC on 23 December 1912:

> You must become a Capital City, not only in name, but in fact; you must make your town a model of municipal administration; your institutions, your public buildings, your sanitation, must be an example to the rest of India. To attain these results will demand on your part much sustained effort, and the cultivation of a high sense of public duty. I can promise that the Government of India will be prepared to sustain you in those efforts by every means in its power. We shall not forget, when building a New Delhi outside your walls, that there exists an Old Delhi besides us which claims our interest and our assistance.[1]

The short-lived struggle against forgetting was launched by Geoffrey de Montmorency, who was in charge of the capital transfer programme. On 19 March 1912, de Montmorency addressed the question of Delhi's expansion, mainly with regard to protecting the government from charges that the capital transfer checked the commercial expansion of Old Delhi.[2] An emphasis was retained on maintaining the open space that separated the two cities and thus focused on establishing the boundary of the imperial capital. Census data showed the city expanding at between 20,000 and 24,000 people a decade, so a (constant) rate of 25,000 was assumed for the future (the actual increase between 1921 and 1931 was recorded in the later census as 100,605). The report went on to pinpoint the geographical constriction that would so debilitate

Delhi in the following years. Expansion to the north was outlawed because it would enclose the (European centred) Civil Lines with an 'Indian town' that, it was claimed, would also pollute the water supply. To the north-east was valuable garden land and the poorly developed Sabzi Mandi suburb, while to the west lay Sadar Bazar suburb, to the south the new city, and the river to the east. The only option was to claim the land to the south and force the capital to retreat towards Raisina. This new land would be acquired, 'town planned' and the costs recouped with interest. As de Montmorency put it, 'This alternative is the bolder, but the wiser for Delhi.'[3]

However, despite Hardinge's backing, these plans immediately met with resistance upon the financial grounds that would dog the new city's extension throughout. R.H. Craddock insisted that such development be charged to the imperial enclave (i.e. Delhi Province) not the new Imperial capital, to be financed by loans and grants. De Montmorency replied in May 1912 that many of the as yet unconfirmed plans for Delhi were not leaving reasonable space for expansion, but that the decision on the form of the capital remained to be taken.

In this time of indecision over Delhi's future, the Deputy Commissioner Lt. Col. H.C. Beaden wrote to the Chief Commissioner explaining the increasing pressure on Delhi's urban form. The letter, from 1913, argued that an Improvement Trust must be created in Delhi 'before many years are passed', while the pressure on the city brought by the capital meant that it could legitimately claim aid. The city had been provided with no orderly expansion, people had been 'huddled into a totally insufficient area' leading to encroachments, slums and inflation that made building impossible. Beadon's solution was clear: 'The most pressing need of the Delhi citizen is room: room to build, room to work, room to play, room to walk, room to drive, ROOM' (quoted in Pershad, 1921: 200).

The 1913 New Capital Project Proposal that was finally agreed upon ignored these pleas and demarcated New Delhi's territory at the 500 yard *glacis* around Old Delhi. This left very little space for growth besides the 'Western Extension', the sorry development of which is the subject of the third section of this chapter. The project also included schemes to introduce a water-borne sewage system to Old Delhi, in the spirit of which Sir Malcolm Hailey proposed a small, self-funding Improvement Trust; neither materialised.[4]

The two main explanatory factors behind Old Delhi's failure to deal with its population explosion were, first, historical and administrative and, second, medical and concerned with the administration's inability to accurately survey and 'see' its territory and people. In terms of the former, the Delhi Administration and the Government of India had been in a deadlock over the development of local land since 1874 when the Government of India transferred the Nazul land it had acquired in the 1850s to the DMC.

By 1908, Delhi's Deputy Commissioner admitted that these lands had been mis-managed but left their administration unaltered, despite the advice of the Settlement Officer who stressed the need for a 'comprehensive scheme' of improvement.[5] When the Deputy Commissioner took charge of the lands in 1925, both the central government and the DMC refused to pay for their improvement, causing most of the small schemes in progress to stall.

Second, the Delhi Administration proved inept at tracking the developments within its territories. Governmental rationalities depended upon knowledge about their territories and, thus, upon ways of organising and collating that knowledge. The development of statistical techniques in the nineteenth century allowed the vast variety of a population to be simplified into trends, patterns, densities and, thus, policies. When applied to cartography, these non-topographic maps marked '... perhaps the clearest example of the map in the service of a liberal form of governmentality' (Joyce, 2003: 51). However, Matthew Hannah (2000) has shown that there were certain requirements and processes that were necessary to guarantee that statistics and mapping secured this level of functional efficiency. A sufficient infrastructure had to exist to enable the 'abstraction' by which the complex world became accessible. Second, this world had to be subject to an efficient process of 'assortment' such that it was known through rigorous and reliable categories. Third, the information had to be 'centralized' and analysed by an active and efficient state. Old Delhi in the early twentieth century displayed none of these features.

In terms of cartography, in 1907, the Delhi Administration was still using maps from the mid-1880s despite the rapid expansion of the suburbs outside the city walls. The incredibly detailed Wilson map of 1908 absorbed so much money that there were no funds left for the sanitary works it had set out to enable (Hosagrahar, 2005: 139), while the mapping failed to induce an accurate cartographic mindset in the local administration. When Beadon, the Deputy Commissioner, addressed the need for town planning in 1912, his recommendations were accompanied by a 'rough sketch' that barely resembled Delhi at all (see Figure 4.1).[6] While local maps were used for census collections and boundary disputes, the vision at the level of central administration and strategy remained myopic until the surveys of 1936.

There was also a degree of statistical blindness amongst the administration in terms of the health of its population. The Annual Sanitary Reports of Delhi Province charted not only a decline in health but also an inability to accurately chart that decline. In 1927, Dr K.S. Sethna, the Medical Health Officer of Delhi, began to tackle the need for action over tuberculosis, which was rapidly increasing and was associated with over-crowding and urban congestion. However, the statistics at Sethna's disposal were claimed to be 'lamentably wrong'.[2] Initial information on causes of death was collected by the sweepers in each *mohalla* (walled community) in the city and passed

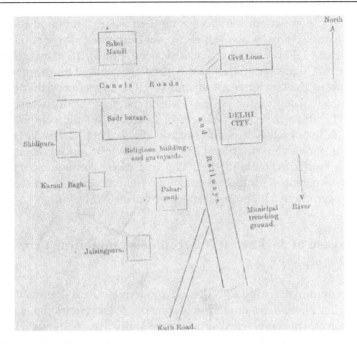

Figure 4.1   1912 sketch map of Delhi. Reproduced with permission of the National Archives, New Delhi

back to their employer, the DMC. The causes listed were usually 'coughs' or 'fevers'. While relatives of the deceased were forced by a local byelaw to notify the authorities of the cause of death, this was often after consultation with local *Hakims* and *Vaids* (Indian medical practitioners) who did not inhabit the same medical episteme as the Western model of the administration. Sethna argued that many deaths categorised as 'other respiratory diseases' (3,237 deaths in 1926) would be tuberculosis, while those recorded as 'Pthisis' (480 deaths) were not actually caused by that disease. Between 34 and 40 per cent of deaths were being listed as from respiratory diseases in the mid-1920s, and Sethna attributed a majority of these to tuberculosis. In the 1928 report it was claimed that 'fevers' were gradually being separated into malaria, the plague, cholera and tuberculosis, yet by 1935 fully 50 per cent of deaths were still undiagnosed.[8] As such, 10,483 deaths were listed as 'other fevers', 1,593 as 'other respiratory causes' and 1,398 as 'other causes'. In addition to this mis-identification, population expansion was adding new statistical problems to that of mis-identification of disease.

The 1929 health report charted the rapid increase in Delhi's population and the spread of *basti* (slums) across the city.[9] The effect on statistics was that birth and death rates were vastly skewed, as these rates included the

new migrant population, but were computed against the population from
the 1921 census. This led to the highest ever birth rate (54.8 per mille) and a
death rate of 47.57 per mille, whereas the death rate for 1930 was confirmed
as the second highest in India.[10] Dr Sethna commented on the report with
the benefit of the new census statistics of 1931. He showed that the rates
had been calculated on the basis of natural increase (births minus deaths)
over 1921, which gave a population of 272,628, yet the census exposed
this data as incorrect to the sum of nearly 75,000 people. The 1931 census
revealed a population of, at least, 347,592 people who were cramming in
to Delhi's limited accommodation. This had severe consequences for local
health, which helped to force a major problematisation of the local and
central governments' biopolitical stance towards Old Delhi.

The disease of darkness: Tuberculosis and failing urban technologies

In his penetrative analysis of French modernity, Paul Rabinow (1989)
claimed that the Parisian cholera epidemic of 1832 set in motion a series of
technological, epistemological and cultural shifts that changed the way the
urban sphere was approached. Statistics on the population were combined
with theories on social ordering to reformulate the topographies of urban disease. Similarly, various diseases had left their imprint on the nature of urban
governance in India, not least the plague (Dossal, 1991; Chandavarkar,
1992). However, there was a much slower response to pulmonary tuberculosis, the respiratory form of the disease that was also referred to as
phthisis or consumption (Hardy, 1993: 212). After Robert Koch's 1882
identification of the tuberculosis as air-borne droplets, the disease came to
be explicitly associated with poor ventilation and congestion in housing,
rather than poverty, climate or race. By the late-nineteenth century, tuberculosis was being addressed predominantly through urban sanitation in the
United States (Craddock, 2000). Intervention was thus recommended to
deal with over-crowding and sanitation, but this rarely materialised for the
urban poor, with their substantial migrant populations.

Margaret Jones (2003) has also detected similar patterns in colonial
Hong Kong. Despite recommendations to adopt a more interventionist
stance in terms of town planning, the difficulty of detecting tuberculosis
and of negotiating funding from the self-financing colony left the disease
untackled by the 1940s. Brenda Yeoh (1996: 94) has shown that there
were efforts to deal with tuberculosis in colonial Singapore before the war,
but that they were underfunded and eventually ineffectual. The Singapore
Improvement Trust was established in 1928, but its attempts to regulate the
urban environment were thwarted by underfunding, curtailed powers and

challenges in the court houses. A similar debate took place in Delhi regarding tuberculosis, over-crowding and government finances, with steps eventually being taken towards addressing the re-housing and improvement questions.

Although tuberculosis had been discovered in India in the 1840s and killed more people than the plague or cholera, it was slow to attract finances for remedial measures (Harrison and Worboys, 1997). Ideas that the disease was hereditary persisted, despite the realisation in the 1890s that it was an infectious, air-borne disease. The difficulty of diagnosis made verification of the rise of tuberculosis difficult to attain, yet by the 1920s it was clear that the disease was becoming a major blight on urban populations. However, as Harrison and Worboys argue, because the disease did not directly effect the government or the economy, it received little attention or funding. This fits into the pattern noted by Lenore Manderson (1996: 12) that medical care in the colonies was first applied to Europeans, then to employed Indians, whereas the majority of the indigenous population received little attention from the state. As a city that was not economically vital to the European population, and one that was thought sufficiently contained not to pose a health threat, the tuberculosis crisis in Old Delhi was slow to receive attention, but did contribute to the eventual re-thinking of the way the city was governed.

In 1924 Muhammad Asaf Ali, who would later represent Delhi in the Legislative Assembly, persuaded the DMC to request the extension of the Town Improvement Act (IV 1922) to Delhi.[11] The request was made by a resolution of 16 September 1925 to which Chief Commissioner Abott reacted enthusiastically. However, the response was delayed for 2 years before being rejected in 1927. In response, on 10 November 1927, the DMC put forward a resolution calling for proposals to combat tuberculosis.[12] Dr Sethna contributed a detailed investigation of the disease, listing the statistical inaccuracies that blighted his investigations, as mentioned above. The recalculated figure suggested that tuberculosis accounted for between three and four of every ten deaths in Delhi, although this figure rose considerably for women confined to the house in the *purdah* system. This was because tuberculosis was not to be explained by climate but by the tubercle bacilli and, thus, by over-crowding. This related to people per room, not per acre, and had been vastly increased due to the splitting up of houses into one-room tenements in response to Delhi's population explosion. The conditions were worsened by Delhi's winding alleys that prevented much aeration of the houses.

Sethna's proposed solutions were for tuberculosis, not over-crowding, and were much more comprehensive than mere town planning: the solution '… should not only comprise the curative side of the problem but should thoroughly involve the various known preventative measures'. Therefore, many of the measures were medical, such as improving health education, establishing dispensaries, hospitals and schools for disease sufferers. Yet, 5 of the 12 recommendations concerned the built environment, including limiting the number of dwellers per room, opening up the slums, expanding the city and improving it within, modifying the building byelaws and providing cheap housing for the poor, on the lines of that in Bombay. Setha concluded by claiming that 'Tuberculosis is a disease of darkness and it casts a blot on the present civilization.'

A year later no action had been taken, and Reverend J.C. Chatterjee raised the issue in the Legislative Assembly, drawing attention to the Assistant Director of Public Health's continued warning about the likely continuance of tuberculosis given Delhi's congested state.[13] A further question in September 1929 pointed out that 418 of the 4,315 female deaths in Delhi during 1928 were from phthisis and the report for 1929 showed that conditions had continued to worsen.[14] The report argued that the conditions in Delhi had become '… most unfavourable to the sustenance of human life', especially for the poor, whose one room tenements decreased the body's natural resistance, as well as increasing the opportunity for spread of infection.[15] Delhi's hemmed in state and its status as capital thus justified a '… systematic, regular, and well chalked out …' system of improvement. Sethna concluded with a 33-point listing of the needs of the city, covering water disposal, sewage, drainage, paving, markets and city planning. In conclusion, he stressed that modern medicine aimed to be preventative as much as curative. Although the government might have a 'sanitary consciousness' what was lacking was a biopolitical 'health atmosphere' in the general public. The government needed to encourage popular cooperation, such that people would responsibly dispose of refuse, use drains and not hide infectious diseases. However, this call for a more interventionist, welfare-oriented state fell on deaf ears.

The extension of the Town Improvement Act to Delhi had been denied purely on the grounds of finance, so cheaper, less comprehensive, measures were decided upon.[16] Stow made this decision following a letter from the Secretary of the Education, Health and Lands Department of the central government, received in May 1926. This stated that, due to improved financial circumstances, the government would now consider funding public health initiatives in centrally funded areas like Delhi.[1] As such, it vowed to consider 5-year schemes, starting from 1927 to 1928. The DMC advised the Chief Commissioner on an ambitious programme that would have cost Rs 6,395,000, although Stow only petitioned the government for a quarter of these funds. The proposed schemes dealt with drainage, hospitals, dispensaries and small-scale alterations to the urban landscape. However, the Chief Medical Officer reminded the Chief Commissioner, in January 1927, that Delhi also had duties to fulfil as an 'imperial capital'. As such, a hospital was proposed on the land to the south of the walled city, where the DMC had been petitioning for expansion of the city to relieve pressure on the slums.

In considering how to administer the Delhi Improvement Scheme's money, the Chief Commissioner consulted Rai Bahadur Mr Sohan Lal, the Secretary of the DMC, who produced a list of recommendations in March 1927. The Deputy Commissioner responded to these points, agreeing that the city needed extension and inducements to get people out, rather than attempting to improve the walled city itself. Despite enthusiasm shown for extensive works on slum demolition in the south of the city and expansion beyond the city walls into the *cordon sanitaire*, the 1929 health report showed that plans to extend the city had been blocked, with the land being given instead to the Government of India Press.

The schemes that actually received the money involved little innovation and were mainly the continuation of existing projects. There was not a lack of money, and there had been some achievements, such as engineering works and some sweeper model dwellings.[3] Similarly, the Burn Bastion and Garstin Bastion road developments around the old western city wall were continued, as were roads stretching out through the western suburbs. However, between 1930 and 1935, as the population continued to increase and the census statistics of 1931 showed just how big an increase had taken place in the 1920s, only two projects actually received any money.[19] One was a road development and one extended electric lights into the Western Extension, which had eventually been foisted onto the DMC. The DMC were still passing resolutions pleading for a city extension, but admitted in June 1934 that the central government had blocked development to the south and east of the city, while the north was too densely occupied and the Western Extension had stalled.[20] The combination of urban congestion, deteriorating health and government inaction attracted criticism from various levels of society, the sum total of which represented a significant and increasingly widespread attack on the regulation of the capital region.

The 'Delhi Death Trap': Problematising urban governance

The 1930s saw pressure grow not just upon the DMC or the Chief Commissioner, but also upon the central government. After the inauguration of New Delhi, the dissociation of the two cities in public opinion became harder to sustain. The criticism against the government came into the civil sphere, via protest movements, the press, and from the DMC as an institution that bridged the public sphere and that of the state. But protests were also made by the Delhi Administration against the central government. What is more, these three levels were indissociable as they fed off each other's protests, information and actions.

While the government *in* New Delhi had angered the DMC by refusing to fund the Western Extension and blocking the expansion of the city south (see the third and fourth sections of this chapter), the government *of*

New Delhi added further insult to injury. In the late 1920s the New Delhi Municipal Committee (NDMC) had begun to criticise its unsanitary neighbour to the north. In 1927, the NDMC complained about the dumping of night soil between the two cities, while in 1929, a sanitary landfill just south of Old Delhi was declared '... a menace to the health of the New City' (Prashad, 2001: 125). The DMC sent a deputation to the Chief Commissioner in October 1929, pleading themselves that the poor health levels in Old Delhi would come to affect the capital.[21] Indeed, Sethna's report on the state of public health in New Delhi for 1930 did little but stress the polluting and pathogenically transgressive nature of its neighbouring city.[22] The sewage arrangements around the railway station near Old Delhi were said to present a '... serious menace to the health of the New City', while the rubbish dumping ground at Paharganj and outside Delhi Gate, to the south of the city, contributed to the fly nuisance plaguing the bordering parts of the capital.

By the time of his 1930 report, Sethna was obviously becoming frustrated with the lack of action over his repeated warnings regarding tuberculosis.[23] He expressed his surprise that people still bothered to ask him why the disease was on the increase and in his 1932 report shifted the tack of his argument from stressing the already terrible conditions in Old Delhi to the potential impact of these conditions on the new city. If housing was not supplied, he warned, '... this fell disease which has now got such a stronghold over the population of Delhi City will spread and thousands will be decimated in the course of a few years'.[24] The report of 1933 argued that this was beginning to occur, after cases of pulmonary tuberculosis had been reported in the capital.[25] However, it was another disease that was the first to cross the *cordon sanitaire* and impact upon the imperial zone. Cerebrospinal fever (a form of meningitis) broke out in Old Delhi in January 1933 and by the end of the year had infected 142 people, killing 68 of them. In addition, there had been two victims in New Delhi, both of whom died. Significantly, the fever was attributed to Old Delhi's congestion with its obstacles to the free movement of air and insufficient sanitation.

The pressure for action increased significantly over the next two years, each case of which drew attention to the dichotomy between the new and old, the safe and unsanitary, cities. The debate was reconfigured on the national scene with the publication in the *Bombay Times* of an article entitled 'Delhi the Death Trap', which was reproduced in Delhi's *Hindustan Times* on 19 January 1934. The article claimed that while Bombay was historically renowned for its slums, '... Delhi is now buttoning on the mantle of disrepute'.[26] Significantly, and accurately, the slums were blamed directly on the lack of cooperation among the DMC, Delhi's local government and the Government of India. Dr Sethna's health reports were used to prove the longevity of the condition and the complaints about it, associated with

a 40 per cent population increase in a decade and an alleged 255 per cent
rise in tuberculosis in just 6 years. Beadon's request for a city extension
18 years previously was also reported. While the people were blamed for
electing sweetmeat sellers and potters to the DMC as part of the Civil Disobedience campaign, the true blame lay with the Government of India for
refusing to extend the sanitary benefits from the New Delhi to the Old.

The article had far-reaching significance. Four days after its publication, the Private Secretary to the Viceroy wrote to the Chief Commissioner
informing him that His Excellency was '... a little upset ...' that his capital
had been labelled a death trap, and inquired with regard to the accuracy
of the reports. The Chief Commissioner had long been trying to mediate a
Municipal Committee pressing for extra funding and a central government
refusing to give it. Plans for the improvement schemes had come to nought
and in 1927 Stow was reducing to going to the home of an Additional Secretary to the Education, Health and Lands Department in order to plead the
case for funding.[27] In considering his reply to the Viceroy, Chief Commission Johnson recounted the efforts of the DMC but laid down his thoughts
on 23 January 1934 in a handwritten, confidential note for file:

> I am afraid that the GoI's [Government of India's] policy as regards Delhi
> city and the development of Government lands is undoubtfully open to attack.
> Delhi – although so important a part of the complete urban area – has been
> neglected for New Delhi, and so often one finds the view advocated on Sectt.
> [Secretariat] files that Delhi has profited vastly by the arrival of the GoI, and
> that it is the Delhi Municipality – not Government – which should find more
> money ...

The Administration's written response to the newspaper article simply highlighted a series of schemes that were proposed, suggested, considered or
tentatively underway. The response was composed by the Chief Commission and, presumably in a show of solidarity, Rai Bahadur Mr Sohan Lal,
the Secretary of the DMC. The article listed the progress made in the area,
although they actually referred to heath policies rather than tackling the
slums themselves. However, the article then went on to launch a thinly
veiled, and at times outright, attack on the policy of the central government.
The central government or private land owners owned all the areas listed in
the article as being particularly mis-managed, while the others were either
very small or awaiting government funding. As such, the Government of
India was shown to have failed in carrying out the ordinary duties of a landlord, and sovereign, and thus has to take its share of the blame for the Delhi
situation. In a stark reference to the uneven conditions in the two cities, the
response concluded that 'The writer forgets that it is easy to build a new city
but it is very difficult to improve an old one.'

The loyalist *Statesman* newspaper, also printed in Delhi launched a further attack on 4 December 1934. Dr Sethna's comments were used, again, to emphasise the alleged difference between the NDMC area, with 2,044 people per square mile, and the DMC, with 58,273. The rampant diseases in the latter made comment on the contrast between the two areas seem superfluous. The need for an improvement trust was clear, the article claimed. The distinction between New and Old Delhi was voiced again in protests on 30 August 1935. In demonstrations at the Town Hall, protestors stated that 'New Delhi people have not descended from the heavens ...When the Government can spend millions [on New Delhi people] why should it treat the Delhi people as untouchables?' (Prashad, 2001: 131). They went on to suggest that all the congestion and disease in Delhi was due to the transfer of the capital from Calcutta.

The difference between the two cities continued to be made apparent in the health reports which also stressed the risk to the capital posed by the congested city to the north. In 1935 it was made clear that the death rate amongst Hindus was 32.13 per mille, in Muslims it was 30.83, but amongst Christians, most of whom lived in either the Civil Lines or New Delhi, it was just 10.78.[28] Despite this, Old Delhi was still posed as a threatening presence. Enteric fever was said to be spreading into the capital due to the flies around the dumping ground inbetween the two cities, although New Delhi recorded only 4 deaths, against 599 in the older city. The incidence of cerebrospinal fever had also increased to 15 deaths in New Delhi, against 216 in Old Delhi, while the recorded deaths in Old Delhi from tuberculosis had increased to 1,013, from 977 in 1935. This was claimed to be having repercussions on the health of the capital, although New and Old Delhi spent the same on health measures despite having completely disproportionate populations.

To add to the pressure from local government, municipality and civil sphere, questions were also being raised in the Legislative Assembly. In addition to the constant barrage of questions about individual schemes, on 12 September 1935, Muhammad Asaf Ali followed up on his tireless campaigning for an improvement trust.[29] Asking when the government would hand over the Nazul land for development in order to aid congestion, the government replied that an officer had been placed on special duty to ascertain the degree and nature of congestion, the success so far achieved in tackling it, what should be done to remove congestion, and the way such a body would be funded. The report was completed in under a year and led to the establishment of the DIT.

Delhi joined a growing list of Indian cities that had needed institutional action to halt the deleterious effects of *laissez faire* policies on the urban environment. Larger cities had long been experimenting with urban development throughout the nineteenth century in, for instance

(Singh, 1979: 25):

- 1794 Calcutta: Improvement Authority
- 1803 Calcutta: Town Improvement Committee
- 1847 Calcutta: Town Improvement Board
- 1864 Bombay and Madras: Sanitary Commissions
- 1898 Bombay: City Improvement Trust
- 1903 Mysore: Improvement Act
- 1911 Calcutta: Improvement Act
- 1919 United Provinces: Improvement Act
- 1920 Andra Pradesh: Town Planning Act
- 1920 Rangoon: Development Trust Act
- 1922 Punjab: Town Improvement Act
- 1926 Nagpur: Improvement Act.

Each of these Improvement Trusts had features in common, drawing on international discourses of town planning and the historical experiences of slum demolition in the west. However, each Trust also had to contend with its local social, economic and political context. Sandip Hazareesingh (2001) has shown how the activities of the Bombay Trust took place amongst class-driven processes of urban development that privileged profit over improvement. Similarly, Siddhartha Raychaudhuri (2001) has argued that urban development was not simply a process of colonial ordering, but also one of negotiation with indigenous elites and the municipal committee. Again, Nandini Gooptu (2001: 91) has shown how Trusts throughout Uttar Pradesh worked to segregate the rich and poor in order to protect the 'superior residences'.

In Delhi, the primary context was that of the political capital and a tightly financed local administration. While the Trust was given the apparatus, image and agenda of a freely operating agent in urban reform, it was actually operating in a tightly structured space of financial, aesthetic and political considerations. These structures were not just deeply causal and did not simply operate through the reified free hand of the market. These were structures that were very consciously debated well in advance of the Trust's formation and were constantly attuned and readjusted so as to prevent the outflow of finances and control from the new to the old city.

Therefore, this section has shown that in Delhi the domain of the economic had, in most instances, been prioritised over the domain of the biopolitical. While this could also be viewed as a protection of the social, articulating the fear that urban clearance schemes would lead to political unrest, the economic has been shown to be dominant in two senses. First, the Delhi Administration's ability to form governmentalities of urban health was limited by underinvestment in the technologies of visualisation that would have allowed the problem to be diagnosed and monitored. Second, when the scale of the problem became clear, there was a steadfast reluctance to invest, financially, in securing the biopolitical process of population reproduction. Only when the political context became threatening was the problematisation deemed a sufficient crisis to launch the governmental machine into action. The following sections not only trace the emergence and transformation of this machine, but also display the continuities in how the technologies and visions were limited by an ethos that distributed resources towards the new, not old, city.

Congestion Relief, Calculation and the 'Intensity Map'

Visualisation, the intensity map and the imperial ethos

On 3 August 1935, Andrew Parke Hume wrote a letter to his father, informing him that he had been requested by the Government of India to investigate the relief of congestion in Delhi.[30] The letter set out the tasks that would, a month later, be announced in the Legislative Assembly as the duties of the newly appointed special officer. What this statement did not include was the rider at the end of the letter addressed to Hume. It stated that there was no budgetary provision to fund the research, that a supplementary grant was undesirable and that cuts should be made in the existing budget of the Chief Commissioner to fund the investigation. Hume referred to the letter as 'beastly', as it would mean him missing his leave in England; he also had a plethora of other equally negative adjectives that he would apply to the Government and their financial strictures over the coming years (see Legg, forthcoming 2008, for a discussion of Hume's correspondence).

Within the intensely political landscape of Delhi, Hume's views were often articulated in relation to the juxtaposition of the two Delhis. After his first meeting with the Chief Commissioner, he wrote home that there had been a mass of suggestions for improvement over the last 15 years, '… but Govt. have never followed any definite policy with regard to Old Delhi, lavishing all their interest and money on New Delhi'.[31] Similarly, in a letter written on 18 and 19 August, he claimed that '… Old Delhi has waited over 20 years for the Govt. of India to take pity on its squalor and slums whi[le] they poured out their gold on the Imperial New Capital'. He was equally scathing about past efforts by the government to avoid spending money or using up its land in Old Delhi, calling it 'appalling' that only a few plots had been 'fiddled around with', and insisting that effective action would cost a lot of money.[32] He suggested as much in what he described as the 'battle royal' at Simla, 'the abode of the Gods'.[33] At the meeting in mid-September, he informed the heads of various central government departments that he would have to act on a large scale and with lots of money. Over the following months he planned how he would do this but, suitably forewarned, the Government of India also began its own preparations.

During his investigations, Hume would come up again and again against financial restraints imposed from above. On 16 February 1936 he wrote of his attempts to milk 'the governmental cow' for Rs 2,400 for an aerial photo of Delhi, and told of his disgust, on being denied, that, 'When it comes to New Delhi we don't talk in terms of "thousands", but in "crores" [tens of thousands].' However, his report was completed and set not only a new form of practical improvement in Delhi, but also a new way of envisaging the city that would infuse the practice of improvement itself.

Hume did not face his task alone. He was joined by two of the most indefatigable proponents of urban reform in the last decade. Dr Sethna and Rai Bahadur Mr Sohan Lal assisted him in his investigations and, in his diary, Hume noted the work they had done on his behalf during his leave in England during the autumn of 1935.[34] This consisted of most of the statistical calculations of congestion on which the report was based. The report was a response to the impenetrable darkness of Delhi in terms of statistical knowledge. Returning to Hannah's (2000) criteria, the report sought to work at a level of 'abstraction' through performing calculations on adjusted census statistics, it 'assorted' the data by an analysis of over-crowding in the city and it 'centralized' this information within an Improvement Trust that would aim to put this statistical knowledge into practice.

This eventual practice had multiple and diverse origins. Hume visited various county councils in the United Kingdom to assess town planning practice, and on return to India wrote favourably about the work in Bombay and Calcutta.[35] The Viceroy himself suggested that an improvement trust should be an outcome of the report, which while under preparation had to negotiate the many bodies that had previously managed the landscape and had their own vested interests.[36]

The report was finished on 5 May 1936 and highlighted the prior deadlock between the Government and the DMC over the improvement of Nazul lands.[37] Hume managed to camouflage much of his distaste for the Government's inaction through providing a painfully detailed description of the number of failed attempts to ease Delhi's expansion westward. Despite this, Hume could not refrain from labelling the constant calls for Nazul reform over the years '… almost pathetic …' and from pointing out that the new capital took much Nazul land from DMC control and that, after 1911, 'Town planning of the older city received attention mainly in its relation to the requirements of the New Capital.'[38] Hume recounted the failure of the Western Extension to provide substantial accommodation for Delhi's population expansion, which was exposed at 40 per cent per decade in the 1920s by the 1931 census. Noting that the DMC would refuse to service the Nazul lands they did not own, while the Government

Table 4.1 Census returns: Population increase in Delhi City, 1881–1931

Year	Population	Decadal increase	Percentage increase
1881	173,393		
1891	192,579	19,186	11
1901	208,575	15,996	8
1911	232,837	24,262	12
1921	246,987	14,150	6
1931	347,592	100,605	40

would not give away such lands for free, Hume asserted the urgent need for reform.

In assessing the nature of the congestion problem, Hume focused almost entirely on population description, eschewing an emphasis on disease, economy or the cultural landscape. Chapter II of his report began by pointing out the rapid increase in population expansion in the 1930s (see Table 4.1). The 'patch-work policy of city development' had been unable to cope with this increase, although Hume reneged on his direct criticism of the Government by arguing that Old Delhi had been congested since the days of the Mughal zenith and thus the problem was also one of great antiquity.[39] The problem was thus identified as one of the congestion of people in houses and houses on land within the walled city of Delhi and the surrounding suburbs.

The extent of the former problem was pinpointed by a statistical analysis of the 1931 census data. On the assumption that Old Delhi could support 200 people per acre, Hume calculated that Delhi Municipality had an overpopulation of 88,169, which was rounded up to 100,000 for 1936–7. Hume went further and produced a map setting forth the geography of population congestion in the city. This map was deemed essential because it would allow people a concrete basis on which to estimate the nature of the problem. The 'intensity map' as it was referred to, plotted population density per acre through the city (see Figure 4.2).[40] The 10 most overpopulated circles in terms of *excess* population, indicated in an appendices to the report, were all crammed into the walled city. This map was not simply passive and descriptive. It was used to make the nature of the problem clear and helped define the nature of action that would be necessary. In Hume's words, the object of the Trust would be to 'level out the intensity map'.[41] As such, the sole object of the Trust would be to improve standards in the centre, demolish slums and provide areas for the city to expand outside the limits of 1937.

The problem of houses on land was harder to stress by statistics or maps due to the lack of accurately updated cartography. Hume simply described

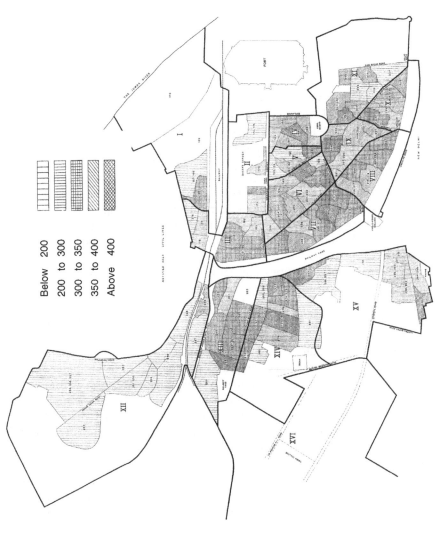

Figure 4.2 Population intensity in Old Delhi, 1931

the mass of small alleys, often overbuilt to render them little more than tunnels, with the residences springing off them being almost entirely deprived of light and air. Ventilating courtyards had been built over, drainage systems had collapsed and slums had emerged not just in the walled city, but also in the surrounding suburbs. This was shown to be having an effect on infant morality and tuberculosis rates, for which Hume referenced Sethna's reports.

Having assessed the schemes already underway in Delhi, Hume concluded that they were completely inadequate. At full development they were estimated to only hold 40,830 people, leaving a shortfall of 59,170 for present needs, while the population was expected to increase by a further 133,000 by 1951. Hume recommended that a special officer be appointed as the head of an Improvement Trust that would combine the Nazul lands with any lands in the process of development by the DMC. This would institute new schemes that Hume listed, most of which featured in the reports that would be submitted for approval after the Trust was founded in 1937. However, the continuity between Hume's proposals and the Trust that was formed should not disguise the governmental manoeuvrings that took place in order to ensure the ethos of the imperial system would not be compromised by Hume's fervent desire to remedy the biopolitical catastrophe of Old Delhi.

Michel Foucault's writings constantly impressed upon us the imbrications of the spoken, the written, the performed and the material. This is whether thought of as the links between primary, secondary and tertiary spatialisations (Philo, 2000), the preferability of general over total histories (Foucault, 1972) or the formative role of institutional spaces on our personalities (Foucault, 1977). The achievements of the Trust depended upon its ability to encourage people to occupy places in a different way, as framed by the new material spaces it would create. Yet, within, and beyond, these social and physical spaces were the financial spaces of the account sheet and the political spaces of the administrative office. It was in these parallel domains that Delhi was re-imagined, not just on the two-dimensional sheet of the intensity map. These spaces surpassed the limits of the walled city and stretched to the very heart of New Delhi. The following discussion will seek to show how the founding and operation of the Trust was consistently infiltrated by the ethos, usually emanating from New Delhi, which protected the interests of the Government of India against undue compromising of its core ideals. These can be simplified into two categories: first, a reluctance to commit substantial finances to un-remunerative projects, which were usually those that would benefit the poorest slum dwellers, and, second, a political–aesthetic impulse that sought to protect New Delhi and to landscape the old city into the image of a capital.

Imperial financing

Hume knew the figure of 100,000 excess dwellers in Old Delhi would make the government 'sit up' and take notice.[42] On 2 August, the Viceroy informed Hume that he liked the report very much and would recommend the establishment of a Trust. The report was, according to Hume, read widely throughout the central government and the Finance Department spent days 'ripping it to shreds' and placing the least optimistic light on it, although approving it in the end.[43] It was under this pressure that Hume made a promise that allowed a way-in for the imperial financiers. He argued that, with common sense, the government could reap a rich profit from the improvement of Delhi, while also benefiting the city. It was upon this assumption that the Viceroy reordered the whole of the Delhi Administration, with the Trust at its heart.

Hume wrote on 24 September 1936 that His Excellency had ideas on revolutionising the running of Delhi, bringing in new commissioners and coordinating the governance of the New and Old Delhi Municipalities. Indeed, the Viceroy issued a minute on the 'Future Administration of Delhi' on 26 September 1936 that had far-reaching consequences for the Trust (see the second section in Chapter 2).[44] Regarding Old Delhi the Viceroy concentrated power in the hands of the new Chief Commissioner E.N. Jenkins, leaving the Deputy Commissioner to deal with law and order. Hume, the Improvement Officer, was not only granted all the powers he requested in the report, but was also named Chairman 'ex-officio of the New Delhi Municipality'. This placed Hume in charge of all matters of improvement between the two cities. Included within this remit was the financing of a Sewage Scheme from the projected *profits of the Trust*. The Government would give a grant for half the cost and a loan for the remainder.[45] Similarly, malarial improvements works would be funded by the DIT, via a loan from the Government.

The significance of this decision is only made clear in mapping the location of these two schemes. The Delhi Sewage Extension Scheme consisted of seepage recorders, overflow work and outfall sewers in New Delhi and works on the New Delhi power house. Other works were in Kilkori, to the south-east of New Delhi, and targeted the disposal works at the river. In 1928 J.N.G. Johnson, then Deputy Commissioner, had produced a 'Scheme for the Future Administration of New Delhi' that set out the NDMC's role in relation to other governmental bodies in the Delhi Province.[46] It was acknowledged here that the NDMC should pay for all sewage work because the DMC, although passing its sewage into the New Delhi system, would be unable to support the massive cost of maintaining the infrastructure and the heavy interest charges on the capital cost. The main question for Government, it was suggested, was '... who will have to bear the brunt of the financial subsidy'.[47] With the formation of the Trust, this dilemma was resolved.

The Anti-malaria Works focused on filling areas of stagnant water and draining off marshy land. The areas for action included the Jumna village to the north-east of Old Delhi, the drain surrounding the old city and a creek in Qudsia Gardens of the Civil Lines. Other works surrounded Metcalf House in the same area, while the majority of works were around Hardinge Avenue near Nizamuddin to the south of New Delhi. This area was still populated with coolie camps left over from the construction of the capital, but the area was also considered dangerously close to the new city. While some of the malaria works around Old Delhi would have benefited the city indirectly, none of the works addressed the sole purpose of the institution by whom they were funded; the removal of congestion in Old Delhi. The repayment of the loans was debited from the Nazul account, thus preventing any profit from this potentially lucrative account going to Trust anti-congestion works in the initial years. Therefore, effectively, these measures lodged in the foundations of the Trust's administration a capital outflow from Old Delhi to the European-based enclaves of the Civil Lines and New Delhi.

While the Viceroy continued to take an interest in the Trust and worked to organise the local administration so that it could best aid Hume in his task, the question of profits would hamper the DIT throughout its operation. Hume soon became aware of this and complained on 15 December 1936 that the sewage and malaria works were 'I think, a mistake, because they're outside legitimate Trust activity and are likely to swamp the other side of this work'.

In addition to the burden of funding the malaria and sewage works, the Trust was also riven by an imposed accounting division between the Nazul land and the Trust Estate. The Nazul Estate was 'placed at the disposal' of the Trust to 'hold and manage', with all improvement costs being borne by the DIT.[48] A condition of this transfer was that separate accounts of revenue and expenditure would be maintained for the Nazul Estate, from which the Trust would pay Rs 200,000 per year to the Government, being the net income of the Nazul Estate for 1935–6. Not only this, but any surplus sum remaining in the Nazul Development Account at the end of each year would be put at the disposal of the Government, who would either order it to be used for Nazul development or for paying back loans made by the Government to the Trust. From this money, the Government gave back a grant of Rs 21.4 lakhs, which was equivalent to half of the cost of the sewage disposal scheme, yet it only made loans available for further work. The consequence of this accounting division was made clear in the debate on the constitution of the Trust that stressed that the net revenues of the Nazul must be devoted to repaying the Government loan, thus refunding all profits to Government.[49]

This infuriated Hume, who wrote on 1 November 1936 of his amazement at the government's irresistible temptation to interfere and make things as complicated as possible. Hume's proposed simple office system of accounts had been interfered with, he claimed, 'for political and other reasons'. As the DIT took form in early 1937, Hume privately reiterated his concerns that the Finance Department had retained too much control and would interfere too much.[50] He made these concerns known in a confidential note on the proposed DIT constitution, made in February 1937.[51] Hume expressed his apprehension over a system that made the Trust dependent on the government granting a loan and wrote that he had hoped for the Nazul proceeds to be placed at the disposal of the DIT. Since the Nazul land would not make a profit while being improved in the first 2 years, paying the Government their full demands would inevitably drain the Trust's activities. Similarly, he stressed that any remaining money would be used on malaria and sewage works that were not central to the improvements '… for which old Delhi city has so long waited …' The effect of the Trust's extra duties would be to '… drain funds otherwise available for Delhi city'.

These tensions lay barely effaced beneath more of the correspondence between Hume and government figures. In response to a question in the Legislative Assembly in January 1937, the central government stated they had not decided whether Hume's report would be made public.[52] The Chief Commissioner asked Hume for his opinion and he replied on 12 March that there were two grounds for not issuing the report; embarrassment to the Government or embarrassment to the Trust.[53] In terms of the latter, the report mentioned the lands the Trust hoped to acquire, which could lead to land speculation. In terms of the former, Hume pointed out that his report had estimated that the Government estate (Nazul land) would yield a profit of Rs 6,600,000 and that these profits could have been devoted primarily to a general scheme of improvement. Similarly, it was observed that some would argue that:

> … Government should have given the Trust a more secure financial basis in the shape of a substantial grant: that they should not have appropriated the whole of the Nazul recurring income to general revenues, and a large part of the profits, to financing sewage and malaria schemes while congestion remains a problem.[54]

In discussing Hume's comments with the Government in New Delhi, the Chief Commissioner admitted that criticism of the Nazul policy was inevitable as support for funding slum clearance would gain as time passed. He also laid bare the extent to which Hume's objectives were distant from those of the local and central administrations, and the degree to which his personal dilemma was acknowledged. In a letter from 13 March

1937 to Yeatts, the Secretary of Education, Health and Lands, Jenkins stated that:

> [o]ne objection to publication which Hume has not considered is that the broadcasting of his personal views which are obviously more 'liberal' than those of Government may be a little embarrassing to him as Chairman. There is not much in this, I admit; but it is easier to carry out a policy of which one does not wholly approve if one's personal views are not known.[55]

Relations continued to deteriorate as the Finance Department proceeded to use its powers to determine how the Trust spent its money. On 15 June 1937 Hume wrote to his parents that 'They are an arid body, the worthy of the G[overnment] of I[ndia] lacking vision, courage, Christianity, and inspiration. I can't think of anything else to say against them.'[56]

In 1938, as part of Hume's draft constitution for the DIT, he placed on record his apprehension to the system of making the Trust dependent on a Government loan.[57] This would, he claimed, hamper work in the first two years and divert funds to loan repayment, as well as to the malaria and sewage works. Hume went on: 'It would be unfortunate if the work of improvement for which old Delhi city has so long waited should once again miscarry, because the development agency has thrust upon it extraneous municipal activities greater than it can bear.'

The Chief Commissioner Jenkins had been appointed to aid the DIT and he *did* work to advance their cause, yet his primary allegiance in the Warrant of Precedence was, of course, to his superiors. However, the strain of negotiating these two causes did begin to tell in the summer of 1937. On 27 July, Yeatts had written to Jenkins offering his help in publicising the work of the Trust, yet on 14 August the DMC passed a resolution protesting against the decision to debit half the cost of the sewage works from the Trust as opposed to Nazul takings.[58] If the government appropriated Nazul profits, and Trust profits were raided in advance how, they asked, could the city be improved? In light of this, Jenkins sent a sober reply to Yeatts on 18 August 1937, although he also included a 'ballyhoo' (praising advertisement) about the Trust that was planted in the *Statesman* newspaper.[59] He stressed that from the mindset of the Finance Department, it was impossible to survey the Trust's policy without concluding that if it did its job, there would be no profits. Cutting to the quick of the emergent tensions between Hume's biopolitical aspirations and the economic considerations of the Government, Jenkins argued that 'It is difficult to square this conclusion with the declared intention to appropriate to Government the ultimate profits on the Nazul Estate'. Indeed, he suggested that any intelligent critic would point out (as Hume had done) that the Government should have given a Rs 100,000 grant 15–20 years ago, the least they could do now was to abandon the profits of the Nazul. Referring to his ballyhoo, Jenkins argued that 'I have of course dealt in the article with the broad aspects of Trust policy, and it is quite likely that no one here will be intelligent enough to discover the joint in our armour, of which I am acutely conscious.' That is, that the Trust would be struggling to pay off its initial burdens for many years and would raise no profits.

Jenkins continued to make this point in May 1938 in response to the Viceroy's enquiries about the Trust activities.[60] Jenkins replied to the Private Secretary to the Viceroy, reminding him of the decision to debit half the sewage works costs and all of the malaria costs from the Trust. Jenkins continued: 'I have always had the gravest doubts about the suitability of these arrangements. In the first place, City improvement cannot on a long view be remunerative, and the profit motive underlying the separation of Nazul account operations is therefore misconceived – that is, unless the primary object is profit rather than improvement.' Without the Nazul's income, the Trust had to pay to acquire land on top of its 10.5 lakh unremunerative debt. As such, Jenkins was at a loss to see how any work without central funding could be done without making a loss.

Imperial aesthetics

The second set of strictures imposed upon the Trust expressed the imperial ethos through notions of aesthetic landscaping. The mounting critique of the management of Old Delhi throughout the 1920s and 1930s had mobilised the idea of 'Delhi as capital' to the call of urban reform. This was a tactic Hume had taken up in his report and used to further his cause, as will be shown in the discussion of the schemes. The idea of such a congested city bordering New Delhi was not just posed as a health risk, but also as a threat to the appearance of the capital region. As with the suggestion that the Trust could reap a 'rich profit', the suggestion that it could play a role in making Old Delhi look like part of the capital came back to haunt the DIT.

On 5 May 1938, the Viceroy's Private Secretary wrote to the Chief Commissioner, following His Excellency's tour of some of the Trust's early schemes.[61] Information was requested regarding the steps taken to preserve the vistas and approaches to historical monuments, to protect tombs and the city wall and to protect open spaces around objects of historical interest. It was the Viceroy's opinion that the Government's actions should be reproached for being '... actuated solely by hygienic or commercial considerations ...' and asked that Jenkins '... ensure that the artistic and aesthetic considerations involved are not lost sight of in planning and reconstruction schemes'. To the suggestion that financial considerations should not hamper these points, Jenkins sent an exasperated reply reminding the Secretary of the DIT's heavy opening debt. In relation to aesthetics, the main

concern was the Delhi–Ajmeri Gate Slum Clearance Scheme, which provoked a vociferous debate about the political–aesthetic landscape, as shall
be recounted in full in the fourth section in this chapter. Smaller considerations of the capital's aesthetic did, however, pervade the technical work of
the Trust on the ground.

The improvement techne

The government of colonial India was as much a technological task as it was
a political one. Gyan Prakash (1999: 3) has placed science and technology,
as physical structures and administrative regulations, firmly at the heart of
rule in colonial India. The census, surveys, encyclopaedias and other forms
of classificatory information depicted India as a unified and knowable space.
After the Revolt of 1857, irrigation, telegraph and rail networks were cast
across the subcontinent to tighten rule. As such, modern institutions, knowledge and practices *assembled* the Indian nation as a coherent idea and space.
The state was thus inseparable from the technological configuration of the
territory and the modern India it was engineering into existence (Prakash,
1999: 160). However, these technologies were not independent of individual
action but relied upon people knowing and living through their machines,
laws, rules and limitations.

Just a week after accepting his post as special officer, Hume realised that
'It is so largely a technical job for which I have no technical qualifications
and I shall have to get much estimating and details done by the PWD [Pub-
lic Works Department], screwing the work out often by tact and guile.'[62]
Hume attempted to negotiate the legal and technical terrain of the Trust
activities as well as he could, but the financial strictures impeded many of
his grander visions. The DIT came into being in 1937 with the extension to
Delhi of elements of the United Provinces Act VIII of 1919, the Rangoon
Development Trust Act of 1920 and Section 78 of the Calcutta Improvement Act of 1911.[63] The Government would determine the extent of the
Trusts activities, but it could automatically work within the entire limits of
the DMC. The basis of the Trust was the 'Government Estate'; the 23 plots
of Nazul land (the Nazul account) which, it was made clear, the Government could resume possession of at any time.[64] The separately accounted
'Trust Estate' consisted of some small DMC lands given to the Trust but was
mostly constituted by privately acquired lands. As such, the Trust account
had to raise money to both purchase and develop its land.

The legal foundations provided the Trust not only with many powers of
its own, but it also took up existing legal powers and sought to improve
their effectiveness. A series of Municipal Byelaws had been applied to Delhi
in 1915 but, Hume had commented in his 1936 report, it was exceptional

for any house in the city to follow these regulations. The laws dictated that plans for all new constructions be submitted to the DMC, to ensure they conformed with pre-conceived standards and also to allow the detection of deviations in the future. The other byelaws sought to regulate not just the built form but also the biopolitical spaces of the urban form, from the street down to the individual:

(1) *Streets in DMC notified areas*: No encroachment within seven and a half feet of what the DMC deemed was the centre of the road while no new building could be roofed with inflammable material.[65]

(2) *Houses*: Height regulated by the width of the street, from a street width below eight feet having a maximum house height of 12 feet, to streets over 35 feet wide having houses no higher than one and a half times the street width.

(3) *Drainage*: Through cast iron pipes, while no pipes would pass through interior walls.

(4) *Rooms*:
   (a) Physical space: No room less than 10 ft high, while no 'single storeyed house' could have a courtyard of less that 15 per cent the ground area, increasing by 5 per cent per storey.
   (b) Living space: Every inhabited room needed a window or door one-eighth the size of the total floor area opening onto a space six feet wide and open to the skies or a verandah. Every fireplace needed a chimney made of iron, brick or stone while the floor beneath and around the fire for three feet had to be rendered fireproof.

(5) *Hygiene*:
   (a) No open sewer or drain could run through an inhabited room while every 'latrine, privy or urinal' needed adequate ventilation in the form of an opening at least a foot square to open air.
   (b) Non-water-borne latrines needed a metal or pottery fitting with which to remove solid waste while every latrine, bathroom and cooking place was to be sufficiently drained into a Municipal drain or private cesspool.
   (c) The floor of every latrine or urinal was to be impermeable and to slope to allow easy drainage while the walls to a height of three feet were to be impermeable metal or masonry.

(6) *Conduct*: Toilets should be readily accessible to cleansing and '... when the outer door thereof is open, the seats shall not be visible from the street or other public place'.

These standards were beyond the powers and resources of the DMC to impose upon every household. However, they were used by the Trust to identify an area as insufficient or, in extreme cases, as a slum. The DIT had various techniques laid at its disposal, which addressed different aspects of the urban landscape. These can be categorised as follows:

(1) *Site preparation*: purchasing, demolishing and remodelling structures, laying out property and site planning.
(2) *Infrastructure*: work on streets, drainage, water supply, lighting and open spaces.
(3) *Construction*: providing accommodation and buildings.
(4) *Finance*: advancing money for schemes and selling, letting or exchanging property.[66]

An area could be announced in need of improvement should it be 'too badly arranged' or have 'any other sanitary defects', inline with the byelaws already discussed. The techniques listed above were combined into different assemblages to tackle specific cases of improvement. When such a case was decided upon, occupants on land required by the DIT were given 60 days to lodge an appeal, while the Trust could enter any land to measure, survey, assess or 'to do any other thing'.[67] However, these policing powers were checked by certain provisos; all such entries would be in sunlight hours, having given 24 hours notice in which females could be removed to an area of privacy. The assemblages of techniques included the following 'Schemes', which were referred to as either Nazul (N) or Trust (T):

(1) *Re-building or re-housing*: reservation, re-laying out and demolition of sites, loans for reconstruction.
(2) *(Deferred) street*: improving the appearances and efficiency of a causeway.
(3) *Development*: laying out a street structure and regulating construction.
(4) *Town expansion or housing*: larger scale of a Development Scheme.

The main town expansion schemes were those of the long forestalled Western Section (N7, see the third section of this chapter), the Roshnara Extension (T1) and the Northern City Extension (T2) to the north of Old Delhi (see Figure 4.3). The development schemes were mostly run on Nazul land, in order to make it more profitable. They consisted of the housing estates of Daryaganj South (N1), Ramnagar (N2) and Mondewalan (N5), and the commercial areas of Garstin Bastion Road (N9) and Sabzimandi Fruit and Vegetable Market (N13). Deferred street development also took place on land already owned by the Government, such as Paharganj market (N3), estate (N12) and circus (T6). Besides town expansion, the Trust account was focused on the re-housing projects of the Delhi–Ajmeri Gate Scheme (DAGS) (T3, see the fourth section of this chapter) and Hathi Khana slum clearance (T7), while the Nazul account funded similar schemes, such as that at Ahata Kidara (N11), to enable further commercial developments of its properties.

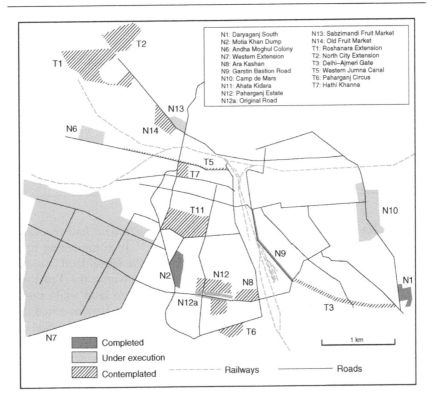

N1: Daryaganj South
N2: Motia Khan Dump
N6: Andha Moghul Colony
N7: Western Extension
N8: Ara Kashan
N9: Garstin Bastion Road
N10: Camp de Mars
N11: Ahata Kidara
N12: Paharganj Estate
N12a: Original Road

N13: Sabzimandi Fruit Market
N14: Old Fruit Market
T1: Roshanara Extension
T2: North City Extension
T3: Delhi–Ajmeri Gate
T5: Western Jumna Canal
T6: Paharganj Circus
T7: Hathi Khanna

Completed
Under execution
Contemplated   ------- Railways   ——— Roads

1 km

Figure 4.3   Trust schemes, 1937–9

Within a year and a half of its operation, Hume wrote to his parents that the international context of the winter of 1938 had made it unsafe to make investments, hitting the Trust's proposals.[68] The approaching war would continue to hamper the improvement proposals, yet Hume continued to apportion blame to the Government of India's lack of vision.[69] On 28 April 1940 Hume admitted that after three years of Trust operations, there was not much to show for the work yet.[70]

This can be explained by the crippling financial burden placed upon the Trust. Hume's first report back to the government for the year 1937–8 pointed out the inherent weakness of the organisation due to its lack of assured income.[71] In April 1938, the Trust anticipated the total cost of its duties for that year as Rs 1.37 lakhs (Rs 137,000) for administration, Rs 45.97 lakhs for Nazul works and Rs 60.95 lakhs for Trust work, but also an extra Rs 42.87 lakhs and Rs 14.78 lakhs for completion of the sewage works and the anti-malaria works, respectively.[72] As such, 35 per cent of the budget was going on these latter projects, forcing the DIT to request a loan of 29.5 lakhs. By 1939 it was obvious to the financial overseers of the Trust

Table 4.2   Expenditure (lakh Rs) from the Trust account, 1936–41

Expenditure account	1936–7	1937–8	1938–9	1939–40	1940–1	Total
Suspense	0	0.22	8.24	27.2	2.68	38.35
Sewage works	0.82	25.26	5.27	−0.2	−0.33	30.83
Trust works	0	1.32	1.75	0.56	9.31	12.94
Surveying	0	0.04	0.006	0.6	9.35	10.01
Loan repayment	0	0	0	3.2	0	3.2
Loan interest	0	0.004	0.6	0.93	0.93	2.47

that the Nazul revenue that was left after the Government had recovered its charge was insufficient to repay the loan debt.[73] In commenting upon this situation, Jenkins noted in October 1939 that the Nazul land could be made more productive and thus service its debt. However, the Trust was in more trouble, having to acquire massive portions of land while simultaneously paying back its loans and the interest on them. Jenkins noted that the Trust schemes were actually those that were planning for large-scale city expansion but that 'I can at the moment see little day light here.' His recommendation was that these schemes be postponed until the financial situation was resolved, and this was largely what happened. Table 4.2 illustrates how disproportionately the Trust account was skewed towards the New Delhi oriented sewage works (minus the Nazul payment) over the Old Delhi oriented Trust works. While the expenditure on surveys and works for the Trust accelerated after the end of the sewage works in 1940, most of the finances after this point were locked up in the suspense account (see Table 4.2). The insecure economic climate induced by the war led to blocks on capital outlays with the money being suspended in Treasury Bills for a safer investment climate in the future. Already owning the land it needed to improve, the Nazul Estate was not hampered by this restriction.

By the end of 1939, no Trust account scheme had been completed yet several Nazul projects had taken formation. Inline with the New Delhi ethos, these were mostly either remunerative or aesthetically improved areas of importance for the capital. The 'Daryaganj South' (N1) housing and street development was completed, which had been described as lying '… on the side of one of the main thoroughfares between Delhi and New Delhi, [and presenting] a sordid and unkempt appearance'.[74]

Motia Khan Dump (N2) had been grassed over for its 'aesthetic value and effect on the health and comfort of the population of Paharganj [which] is not to be reckoned in terms of money'.[75] Following the fears about transmission of disease between the two cities, and especially through the connecting

Table 4.3 Expenditure (lakh Rs) from the Nazul account, 1936–41

Expenditure account	1936–7	1937–8	1938–9	1939–40	1940–1	Total
Western Extension	0	4.5	14.06	8.73	1.05	28.34
Anti-malaria works	0.62	7.52	2.92	0.58	−0.001	11.64
Nazul works	0	1.99	2.89	3.98	1.59	10.45
Sewage works (+ trust payments)	0	4.69	4.69	−0.20	0.08	9.26
Lump sum to government for Nazul land	0	2	2	2.06	2.14	8.2
Interest on loan	0	0.08	1.51975	1.57	1.53	4.7

suburbs, there was a sustained focus on improving Paharganj, which bridged New and Old Delhi. The third Nazul project concerned Paharganj's fuel and *kabari* (furniture) market (N3) and targeted the drainage and sewers of the area. The Garstin Bastion Road Scheme (N9) completed a project begun in 1913 with the demolition of the western portion of the city wall. The area provided a profitable and presentable vista to the train track by which most people entered Delhi and framed the New Delhi Railway Station. A similar frontage exercise was partly completed through the heart of Paharganj with the Original Road scheme (N12). The other completed schemes, such as that at Ramnagar (N4), provided valuable plots that were sold off for a profit.

A similar pattern was detected in the projects still underway. The Champ de Mars project (N10) sought to irrigate and landscape the tourist region around the Red Fort that was later described as '... a drear [*sic*] spectacle of dusty untidiness in the heart of India's capital'.[76] Ara Kishan (N8) slum was demolished and provided with new services, while reserving a grassed open space opposite New Delhi Railway Station. Thus, by 1938–9, the Nazul account could still pay off Rs 29 lakhs for malaria works and Rs 47 lakhs to the sewage works while retaining Rs 28 lakhs to spend on collective works, besides the Rs 140 lakhs it was pumping into the Western Extension (N7, see the third section of this chapter and Table 4.3). The Trust account, however, was paying Rs 5.27 lakhs to the sewage works, and Rs 82.4 lakhs into the suspense account leaving only Rs 1.75 lakhs for works. This meant that the Roshanara (T1) and Northern City (T2) Extensions could only be planned while the Hathi Khana (T7) slum clearance did not take place.

By 1941, when the DIT's next administration report was filed, there had been relatively little progress due, in part, to the war economy (Figure 4.4). The sewage and malaria works had been paid off by 1940, having dominated

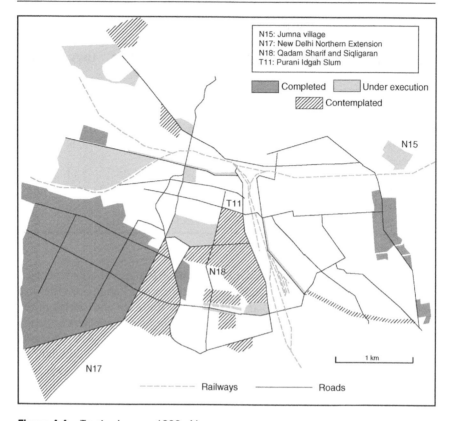

N15: Jumna village
N17: New Delhi Northern Extension
N18: Qadam Sharif and Siqligaran
T11: Purani Idgah Slum

Completed   Under execution
Contemplated

N15

T11

N18

N17

1 km

N17

--------- Railways   ———— Roads

Figure 4.4   Trust schemes, 1939–41

the finances up until this point. The Ahata Kidara (N11) programme was completed, evicting 201 slum dwellers and replacing them with profitable middle-class housing.[77] Besides further work on the maintenance of projects already in progress, the Trust schemes did start to take physical form. Lands were purchased in the city extension zones to the north and basic services laid, although none were completed until 1942–4, after which private contractors had to be encouraged to build on the sites. By 1944 further work had taken place in Paharganj, although this only dealt with the frontage along the streets rather than the areas behind that were 'filthy' and 'wretchedly serviced'.[78] Work was also taking place on the Jumna Village (N15) to the north-east of Old Delhi which was described as a 'disreputable and beggar infested area, which is the first evidence of the Imperial capital to greet the eye of a traveller entering Delhi from the east along the Grand Trunk Road'. Motia Khan slum (T10) was also one of the few to be dealt with, as it was explicitly stated to lie too near the New Delhi border.

Ongoing problematisations

While the criticism of the Delhi–Ajmeri Gate, and Western Extension, Schemes drowned out other levels of critique in the archive, there was a continuous rumble of dissent against the DIT and urban conditions in Delhi. The health reports continued to point out the substandard conditions in Delhi. Reporting in 1937 the new Chief Health Officer, Major W.H. Crichton, stressed that the rise in infant mortality, to 166 per mille, was deplorable, as was 70 per cent of deaths still being from undefined 'other' causes.[79] For the suspected 12,000 sufferers of tuberculosis in Old Delhi, the situations would remain bleak without extra housing. It was acknowledged that the Trust had done some good, but that its failure to plan the building of poor class re-housing would limit its effect. As the nature of the DIT activities became clear, the Health Reports stressed with more urgency the need for slum clearance. Reviewing 1938 it was claimed that without slum clearance measures, '... all others fall into insignificance, if not complete uselessness'.[80] As such, the Trust had not only to clear sites, but also to re-house. Without such measures, tuberculosis would continue to spread, as would diseases like cerebrospinal fever that affected six people in Old Delhi and five in New Delhi that year.

In 1938 Muhammad Asaf Ali also wrote to the government reviewing the attempts to relieve congestion that had been made since he first suggested a Trust 14 years earlier.[81] The Trust and the DMC were equally denounced, although the Government was also criticised, who in '... their pre-occupation with the development of New Delhi and the strict observance of a policy of isolation of the old town... did not even so much as take account of the alarming developments which had begun to make themselves keenly felt by 1923'. The financial division between the Nazul and Trust lands was also declaimed as it forbade the use of Nazul money to fund the Trust schemes, which would be slower to show a profit. Asaf Ali continued to deploy the New Delhi argument, flitting between the risk of the old city and the obligations of the new: 'Having spent many crores [hundreds of thousands of rupees] on the development of the new city the Government should not withhold from old Delhi what is needed for urgent relief and for saving Delhi from becoming a deathtrap and a disease stricken and, therefore, dangerous neighbour of the new city.'[82]

The Trust also came across resistance from the everyday dweller. Petitions were received from 1938 onwards from people affected at the ground level by the works of the DIT. On 10 February a joint residential petition was submitted from the wards of the south and west of the walled city.[83] They requested that land only be acquired, at a fair price, where absolutely necessary for a public purpose and that they be given the chance to rebuild unsatisfactory houses, rather than having them demolished. The farmers of Shahdara complained that the land they had been evicted from remained unused, while they were impoverished. Having had their request turned down by the Trust, they wrote directly to the Chief Commissioner, begging the 'condescension' of his 'magnanimous self'.[84] Further complaints on 30 May 1938 told of people refusing to move from land without being given a replacement plot, one of whom was charged Rs 70 in court over land originally worth only four rupees. Alternatively, another dweller of Andha Mughal was compensated only Rs 93 for land on which he claimed to have spent Rs 700.

The pervasive displeasure found an outlet in November 1938 through the Workers' League, established with the cooperation of local socialist and Congress socialist activists. In a note letter-headed with 'Works of the World Unite', the Chief Commissioner was informed of the resolution of a mass meeting held in Gandhi Grounds in the centre of Old Delhi on 6 November. The mass expenditure, mis-appropriation and embezzlement of public funds by the Trust was condemned, and calls were made for the dropping of all schemes. An independent enquiry was demanded and the punishment of those found guilty.

As many of the schemes failed to materialise, the powers the Trust had maintained over the land it hoped to use came to be criticised. The prohibition of building to the north-west of the city in anticipation of town expansion led M.S. Abdullah, a Municipal Commissioner, to write to Hume in March 1941 that the object of town planning was not that '... the present generations may be ruined and they may be forced to contribute everything for the benefit of the coming generations ...'[85] A year before independence, a question was asked in the Legislative Assembly relating the achievements of the Trust back to its founding financial strictures. The question requested a statement of the DIT's net finances over its period of operation, whether it was true that the Trust had opened with a loan of Rs 35 lakhs for sewage and malaria works that were not its concern, and whether it '... has done practically nothing to remove the congestion of the city and carry out other improvements'.[86] The Government confirmed that the Trust had actually spent Rs 51 lakhs on the sewage and malaria works, funded by a 20 lakh grant and a 50 lakh loan. However, the breakdown of the finances showed that of the Trust's total expenditure, 51 per cent had gone on non-congestion related costs. These included payment to government of pre-Nazul revenue (7.1 per cent); repayment of loans (14.8 per cent); anti-malaria works (7.3 per cent); sewage disposal works (4.9 per cent); and the re-housing loan established as part of the Delhi-Ajmeri Gate Scheme (17 per cent). Of the 42 per cent of expenditure that went on works and schemes, a large proportion of this went into attempts to resolve the crisis that had engulfed the Western Extension.

The Western Extension, Protest and Failed Relief

Administrative deadlock and the call for expansion: 1912–36

> If old Delhi is ever to be anything but a bye-word, [then its] town-planning
> is just as important and just as urgent as the town-planning of New Delhi.
> Until people begin to move out of Delhi to more healthy and airy sub-
> urbs, it is quite useless trying to do anything serious towards improving the
> city ...   (Major H.C. Beadon, 1913, in Parsons, 1926: 201)

A key component to the congestion crisis in Delhi, besides the underbuild-
ing in New Delhi (see Chapter 2) was the geographical constriction of the
city. As the first section showed, the need for room to expand had been
made clear as soon as the capital transfer was announced. The Delhi Town
Planning Committee (DTPC) took heed of these warnings and established
the Western Extension, an application of the town planning regulations
that underwrote the planning of New Delhi. However, decades of mis-
management and shirking of governmental responsibility squandered this
opportunity. The Improvement Trust inherited this project and, as shown
below, spent significantly more on it than any other project. Despite this, the
Western Extension remained mired in various controversies, the evolution
of which is detailed below. This will require a re-treading of the historical
ground to suitably frame the Trust's negotiation of the ongoing project.

The last section suggested that the activities of the DIT were delimited by
two strictures of the imperial ethos, one manifesting itself in financial edicts,
the other in politico-aesthetic imperatives. These strictures are perhaps most
evident in the Western Extension scheme and DAGS. The latter took place
in the symbolic battlefield that separated the two cities, and illustrated the
landscaping ethic as well as the underfunding of projects for the poor, as
shown in the fourth section. The former was dominated by the financial
restrictions and represented an ongoing problematisation of governmental
practice. While Prakash (1999: 13) has shown that technology was essential to the colonial project, he has also suggested that the colonies were
overextended and underfunded laboratories of modernity. Yet, the Western Extension shows that even when money was poured into a project (see
Table 4.4), mis-administration and political imperatives could still thwart
the attainment of improvement.

Until 1893, the history of Delhi's extra-mural expansion (beyond the
city wall) was defined by the Government's fear of 'bursting the walls'
and of maintaining a 500-yard military *glacis* of land around the city
(Parsons, 1926: 64). The construction of the railway directly through this
frozen land meant that the latter policy was undermined to an extent,
while commercial pressure on Delhi as a trade centre increased the calls

Table 4.4 Accumulated Trust expenditure, 1936–41

Account	Trust/Nazul account	Expenditure (lakh Rs)
Suspense	Trust	38.34
Sewage works	Trust	30.83
Western Extension	Nazul	28.34
Trust works	Trust	12.93
Anti-malaria works	Nazul	11.64
Nazul works	Nazul	10.44
Surveying	Trust	10.01
Sewage works	Nazul	9.26
Lump sum to government for Nazul land	Nazul	8.20

for the city to expand. In 1889 Deputy Commissioner Robert Clarke addressed the DMC on the need to systematically plan and construct a new quarter to the city, and managed to push through improvements to Sadar Bazar and Lahori Gate that joined the city with its suburbs (Hosagrahar, 2005: 109).

As shown in the first section, in 1912, Geoffrey de Montmorency argued the case for city expansion as part of the new capital project. The Western Extension was the outcome of his arguments but, in line with Craddock's 1912 insistence that the Government not fund it entirely, it was administered locally.[87] The DMC was given the area, to the west of Sadar Bazar, to administer, while the central government funded the initial development that was estimated to cost Rs 6 lakhs. The Government had blocked Montmorency's proposed Improvement Trust, but did initially provide the DMC with funds for smaller improvement schemes. The area was laid out in a grid pattern and was originally intended for poorer class families displaced by the new capital project.[88]

However, by 1924, financial problems had emerged regarding the progress of the Western Extension. As would happen with the Trust funds, the Government had insisted that the DMC maintain separate accounts for any extension work.[89] This was so the Government could be sure that its finances were being used for city extension, not ordinary administration. The Municipal Committee, it was suggested, had used Rs 3.9 lakhs for ordinary administration in the area, when this should have come from Municipal taxation, and had not spent Rs 5.9 lakhs of the Government grant. This prompted an investigation by the Chief Commissioner A.M. Stow, who argued that the Government was losing too much revenue through having other Nazul lands administered and taxed by the DMC. The Government thus granted control of the lands to the Deputy Commissioner, although a compensatory grant was organised for the Municipality's lost income and for the costs it would continue to incur while servicing the Nazul lands. However, this grant was progressively diminished, despite growing criticism of the extension's failure to materialise.

While Stow appeared to be mostly critical of the DMC, in his confidential notes he also acknowledged the culpability of the central government. When the Government offered Delhi financial grants in 1926 for urban improvement, the internal discussion highlighted the centrality of the city extension. The Chief Commissioner noted in March 1927 that the Western Extension should get drainage, water supply and transport links to Old Delhi without delay. He went on: 'That these essential services have not been provided is due mainly to the preoccupation of the officers in charge of sanitary service in the completion of New Delhi. It is however practically certain that the scheme will not be completed in a reasonable time without considerable financial assistance from Government.'[90] The Viceroy read the notes made on this discussion and expressed his interest in Delhi's improvement and, especially, the Extension. In a note from his Private Secretary to Delhi's Chief Engineer, the Education, Health and Lands Minister, and the Finance Minister, the Viceroy claimed in March 1927 that '... as the congestion in the City is largely due to the fact that the Government of India has descended upon Delhi City, and thereby increased its population, it is up to the Government of India to do all they can to relieve the congestion'.[91] In August 1927 Chief Commissioner Stow personally went to plead the case for the Western Extension in New Delhi, arguing that it should be considered separately to the Improvement Schemes that were being delayed by the central administration. Despite this, Stow was informed that no money could be found for drainage in the Extension in that financial year. In December 1928, the new Chief Commissioner, J.P. Thompson, pleaded again with the Finance Minister. However, the Extension was forced into consideration with the other Improvement Schemes, and it was not one of the few schemes to be prioritised for funding.

In September 1928 a question was raised in the Legislative Assembly regarding the lack of water supply and drainage in 'Karol Bagh', a part of the extension.[92] The Government replied that the Chief Commissioner had applied for funds from the fledging Delhi Improvement Schemes, to which Reverend J.C. Chatterjee angrily responded that the scheme was already 10 years old, and asked when these operations would actually begin? The health report of 1929 stated that the extension had a 'crying need' for a filtered water supply and drainage, the wells having been so overused that their water was becoming polluted and causing severe intestinal disorders in the area.[93] The following health report described the area as still undrained and both untidy and unsanitary.[94] The diminution of the Government grant to the DMC fitted a trend that had seen the Government withdraw from its commitments to funding improvement in Old Delhi. This had led, as Hume reported, to 19,000 people inhabiting the Western Extension by 1936 without essential services and the area being colonised by communities carrying out 'offensive trades', such as hide curing.[95]

In 1931, the DMC had requested that the Government hand over the Western Extension so that the lack of services could be remedied. The Government refused but agreed to undertake expenditure on services in the Extension up to a cost of Rs 10 lakhs. However, by the time the Trust was formed, the cost of these improvements had increased to Rs 23 lakhs and little had been done. In the mean time, the criticism of inaction in terms of city expansion continued. In March 1932 a question in the Legislative Assembly addressed the lack of 'ordinary amenities of life' in the Extension, namely drainage, filtered water, or metalled and levelled roads.[96] The Education, Health and Lands Minister replied that schemes to remedy this situation had been organised, but that financial stringency (the post-Wall Street Crash Depression) meant that the schemes were held 'in abeyance'.

In 1932 the DMC yet again urged action upon the Deputy Commissioner, who also headed the Committee, regarding congestion. Health statistics were used to show that while Old Delhi's death rate was 38 per mille, the death rate of New Delhi was 25 and that of the Civil Lines was 12. The Western Extension's lack of drainage or water was, again, pointed out and attention drawn to the fact that 15,000 people, at least, were forced to dwell in a place without the basic amenities for life. The Chief Commissioner J.N.G. Johnson forwarded the DMC's concerns to the Government and pressed their case, arguing that the lands should be handed over to them, in reversal of Stow's decision of 1924. Johnson pointed out that the Government's pledge to give grants to the Delhi Administration to develop the Extension had been far from fulfilled, in times of both prosperity and adversity. The Government responded by requesting specific information on the area. Johnson's showed that despite only spending Rs 17 lakhs on acquiring the land it had, by 1933, acquired a capital value of Rs 265 lakhs. While Rs 46 lakhs had been spent on improving the land, Rs 44 lakhs had been returned in income, from a population of 19,363. Perhaps unsurprisingly, the Government continued to refuse to hand over such a profitable and valuable enterprise, yet simultaneously refused to grant any extra capital for improvements.

The 'Delhi Death Trap' newspaper article of 1934 drew more attention to the failure of the city extensions. The Deputy Commissioner responded in a confidential note that the Administration was 'stuck' on this topic because of the building of Government quarters on DMC land, which both parties refused to service.[97] This comment also draws attention to the fact that the Extension had undergone a partial transformation of purpose. The need for it to be financially remunerative had led to an increase in middle-class dwellings and fewer poor class dwellings to accommodate those who would, it was envisaged, be forced out of the city by slum clearance. The Extension had also been taken over to an extent by housing for Government Servants employed in New Delhi.

The under-building in the capital project had left the Government with a shortage of houses for Civil Servants and at some point the Extension appears to have been appropriated to fill this need. In 1931 a meeting was held in New Delhi concerning the future development of the city.[98] It was acknowledged that the capital had grown quicker than expected and that town planned extensions would be necessary. However, when the NDDC issued their report in 1939, they showed that no new extension had been constructed.[99] The report claimed that the most revolutionary change in the administration of New Delhi was the decision to use the Western Extension area to relieve housing pressure in the capital, not the congested old city. The writers of the report stated that they did not understand why this change came about, but that the area was built over in 1933 for housing Government and Municipal staff from New Delhi. Therefore, the Western Extension that Hume inherited was one with an intense legacy, one that manifested the administrative deadlock in a scarred and unsanitary landscape. Since Hume placed such emphasis upon 'levelling the intensity map', getting people out of Old Delhi was essential. The Extension showed the greatest potential for this, but its use by government servants and the middle classes, as well as the physical and financial impediments to rapid development, seriously impeded the schemes that depended upon its completion.

Improving the Western Extension: 1937–47

The Trust's first 3-year programme stressed the necessity of accelerating works on the Extension and immediately obtained the Chief Commissioner's approval to do so.[100] By 1939, the Andha Mughal Colony Scheme (N6) had been completed, which was used to accommodate the pig keepers, tanners and 'criminal tribes' that had settled in the city extension area.[101] The Western Extension Scheme (N7) itself was still in progress, although the originally sanctioned works were claimed to have been completed. However, the DMC had forced the Trust to agree to install more drains, while they had agreed to service the completed works.

However, the results on the ground were not as satisfactory as they were made out in the Trust reports. Jenkins, the Chief Commissioner, forwarded comments made by the Chief Health Officer to the Trust in May 1938.[102] Concerns were expressed about the type of private buildings being constructed within the plots laid out by the Trust, which were claimed to be of a 'bad type' and were emerging in areas, which had not been provided with essential services. In July 1938 when Mohammad Asaf Ali wrote to Jenkins reminding him of his efforts to create a Trust in Delhi since 1924, he also argued that '... the western extension scheme of developing the Qarol Bagh Area should not have been allowed to degenerate into the haphazard growth of another slum'.[103]

Some of the complaints of those people forced to live in these conditions have also been maintained by the archive. In the same month as the Health Officer's warnings, a series of petitions were received by the Chief Commissioner from inhabitants of Karol Bagh in the Western Extension.[104] One spoke of roads constructed three feet below the level of the house because a contractor removed the soil to use it elsewhere. As new lanes were added to the road, they were simply dug down to this level, exposing the foundations of the nearby houses and prompting the petitioner to beg the Chief Commissioner to prevent the destruction of his property. In the same month another petitioner began a five-month dialogue with the Chief Commissioner. A road had been constructed two feet beneath the plinth of a house on one side, and two feet above on the other, leading to flooding during the monsoon season. The complainant argued that the scheme was ruining the reputation of the Administration and that only people with friends in Government could get their situation remedied. In August the collective residents of Shidipura in Karol Bagh wrote to the Trust begging that land should not be further subdivided and built upon in their locality as it was actually increasing congestion. As the residents put it, '... we have been passing all this time without any arrangement for light, water or drainage with a hope that sooner or later better facilities will be provided. Probably the idea is based on more commercial consideration than humanitarian or the health's point of view'.

This was, indeed, the case. In October 1939, the Chief Commissioner's Financial Advisor wrote that the Extension works needed to be accelerated and that the middle classes should be induced to move there by cheap transport and facilities for borrowing money.[105] However, the wealthier residents of Delhi, who were now the explicit market for the Extension rather than the poor, would not move into such an isolated and unhealthy place. The DMC had delayed providing sanitary measures and the existing infrastructure was so poor that the Health Report of 1940 labelled Karol Bagh as 'extremely unsatisfactory'.[106] Communal latrines were still being used in many areas, which women and children refused to use and thus were making the homes at least as unsanitary as those in the old city. A series of questions in the Legislative Assembly in 1939 illustrate the reputation the area was acquiring.[107] Attention was drawn to the non-existent sanitary fittings, to the increases in rent, and the inefficiency of the sweepers who had been allotted to service the area. A further question in February 1940 depicted Karol Bagh as being full of lanes and roads that were flooded with water, refuse and rubbish and of containing new houses that were ill ventilated and nothing short of dungeons.[108] Hume's drafted reply admitted that some houses had been condemned as uninhabitable but insisted that the sanitation was rapidly improving.

However, conditions in the Extension continued to deteriorate while criticism of the Government increased. Questions in the Legislative Assembly in November 1940 asked why the area remained so dirty, why nothing was being done to improve the area, why the streets were so dark, and why the DMC was doing nothing to help.[109] The Government responded that most areas were satisfactory and that the DMC had been asked to take over servicing the area and, further, stated that the DMC should not '... be found unmindful of its obligation to this, as well as to other parts of town'. However, by January 1941 the DMC had still not taken over the area and Hume forwarded the Chief Health Officer's notes on 'Qarol Bagh' to the Chief Commissioner.[110] They stated that 'The sanitation of this area is frankly disgraceful and constitutes a serious menace to the public.' Roadsides were described that were choked with earth and debris that collected sullage and were being used as latrines due to the lack of facilities, as were other open spaces as small lanes. Only four latrines were functioning and the Trust was castigated for allowing houses to be built without water closets. The Officer concluded that unless immediate action was taken he would accept no responsibility for the epidemic, presumably of malaria, that would break out during the fly breeding season of the next warm weather. His concerns were not just over public health, but also over '... the scorn and ridicule which we as an Improvement Trust will rightly deserve if it ever became known that we permitted such conditions to exist in an area developed by the Trust'.

After protracted negotiations, on 5 March 1941, the DMC agreed to take over only those parts of the Extension that had been built over.[111] This led to the further deterioration of certain areas such that in 1945 questions were still being asked in the Legislative Assembly that echoed the protests of the former 20 years. The lack of lighting had led to several accidents and thefts and burglaries were said to have increased.[112] The DIT provided the information for the answer, which showed that the DMC had refused to service areas where the house tax did not match the maintenance cost. However, a further question showed that, because of the war economy slowdown, the Committee had offered to consider less built-up areas but that the Trust had failed to respond.

For 33 years the project of expanding Old Delhi westwards had been chronically mis-managed. The imperial ethos had dictated that the extension be done with as little cost, and as little risk to the central government, as possible. The refusal to delegate entire responsibility, and ownership, whether to the Delhi Administration or the Municipal Committee, had led to a refusal to commit sufficient time or resources on all sides. By the time the

Trust was formed, the Extension had a political, material and financial legacy that crippled any substantive attempt at improvement. Despite this, between 1936 and 1941, the Extension took up Rs 283 lakhs worth of Nazul expenditure, as opposed to a combined total of Rs 104 lakhs on all other Nazul schemes and only Rs 233 lakhs on all the schemes of the Trust as a whole.

While the last 6 years of the Trust works were hampered by the war economy, the history of the area and its geographically embedded inequalities suggest that even with more funding an equitable solution would have been hard to find. The structured inequality set up in the accounting sheets of the Government of India became inhabited and traversed by the political agents working for the Trust and the Committee. That a situation so patently created by the central government should have transmuted into a state of near open warfare between these two local bodies marks a missed opportunity to force the Government's hand and demand financial aid. New Delhi was not only allowed to remain aloof, however, but also to colonise the extension to alleviate the stress on its own mis-planned capital. While the Government exerted its ethos from a distance and through financial mechanisms in the Western Extension, it was forced to act more directly in defence of its space in the case of the Delhi–Ajmeri Gate Slum Clearance Scheme. Here, its refusal to fund the re-housing of the poor was combined with its dictates on the politico-aesthetic landscape between the two cities to forestall the second major scheme that could have actively, if expensively, relieved the congestion of Old Delhi.

Slum Clearance and the Strictures of Imperial Finance

The Delhi–Ajmeri Gate Slum Clearance Scheme

In discussing the social technologies of pacification that were developed in mainland France and its colonies, Paul Rabinow (1989: 9) claimed to be examining not discipline and government, or regulation, but discipline and welfare. The welfare state came to embody the full extent of biopower in the twentieth century. It encompassed disciplinary institutions along with more regulative and socially penetrative means of regulation, such as unemployment benefit, insurance, pensions and health care.

While such processes developed in France, Rabinow (1989: 277) emphatically demonstrated their absence in the colonies. Here one saw not the idea of government but devices of social technology. In the colonial context the responsibilities of the state were translated, as stressed in Chapter 1. The state simultaneously functioned at a very close distance, in terms of its ability to use physical force and its expanded powers of surveillance, yet at the same time it was also a very distant state, withdrawing from the irreconcilable 'difference' of the colonial subject (Metcalf, 1994). The sphere of government determined this distance. In matters of policing a close distance was desired, while control was insisted upon in matters of the economy. The social was monitored, in terms of political debate and press discourses, yet general policing and education for the masses were not prioritised. In terms of tuberculosis and other diseases, the colonial state has been shown to be reluctant to intervene, and it is perhaps in these instances that we can see the greatest distance not only between the Raj and its Indian population, but also between the Raj and the welfare state.

This was expressed not only in underfunding (Prakash, 1999; Jones, 2003) but also in spatial distancing (Craddock, 2000: 9). The transmissibility of disease created border anxieties that were epidemiological, racial and political, and thus reinforced segregationist policies of neglect that typify the 'dual city' approach to colonial urbanism. Rabinow (1989: 299) showed how, in twentieth-century Morocco, parks, gardens and European residential zones were planned around the *medina* (walled town), which was surrounded by a *cordon sanitaire*. This led to the over-crowding and 'museumification' of the old cities.

As such, the adaptation of European, liberal, welfare policies to the colonial context can be seen to pivot around two adjustments. First, there was a reluctance to invest and, second, there was a landscaping urge that sought to separate and contain the potentially threatening native population. We can view these elements as expressions of the imperial ethos that seeped into the practical rationalities of government. They were both expressed in the Delhi–Ajmeri Gate Scheme (DAGS), yet this dispute can only be understood in terms of the ongoing debate about the city wall that stretched back to the transferral of territorial sovereignty to the British after the revolt of 1857.

The original city of Shahjahanabad (Old Delhi) was originally surrounded by a massive stone wall that was 8.2 m high, 3.6 m thick, 6 km long and punctuated by 11 major gates. The walls were erected between 1651 and 1658 and helped define the image and strength of the Mughal Emperor's 'sovereign city' (Blake, 1991: 32). The British repaired damage that had been caused to the wall during an earthquake in 1720, which only made retaking the city more difficult during the siege against the Indian 'mutineers' of 1857. After retaking the city the Commander-in-Chief recommended the demolition of the wall in 1861 and only a lack of funds prevented its destruction (Parsons, 1926: 184). Instead, the 500-yard glacis was installed as a military defence, which was originally intended as a temporary measure until the wall had been levelled. As the city expanded, the wall came to hinder development and the Municipal Committee claimed in 1889 that '[t]he only argument in favour of their retention is a sentimental one, and the Committee cannot indulge in sentiment when the material interests of the city are at stake' (Parsons, 1926: 66).

By 1905 no action had been taken regarding the wall so Major Parsons, the Deputy Commissioner, put forward plans to demolish a section from Ajmeri Gate northwards to Kabul Gate. Following the construction of the railway outside the wall, the 500-yard zone, in the words of Parsons, had been shown to be '... an empty word written on waste paper. The existence of these railways make the walls farcical for defensive purposes, as troops would under any circumstances have to hold the railways and not the walls' (Parsons, 1926: 188). It was only after the drawing of the Wilson map in 1912 that Parsons's proposal was accepted and a mercantile boulevard constructed in place of the city wall, which was later known as Burn Bastion Road. The commercial success of this project allowed further developments around the west of the city, including the demolition of the wall between Lahori Gate south to Ajmeri Gate and the construction of Garstin Bastion Road. The logical progression in terms of commercial success and the need for city expansion following these two schemes would have been the continuation of the wall removal south of Ajmeri Gate to Delhi Gate (see Figure 4.5).[113] After 1911, however, this portion of the wall did not separate the old city from the suburbs, but marked the division between the Old Delhi from the capital and, as such, retained its force as an imaginary and military divide

Figure 4.5   Ajmeri Gate

between the two cities. The 1920–30s saw the forces for progress clash against the conservative defenders of the political and aesthetic landscape dividing the two cities.

'A tale of two cities': Delhi's aesthetic and political landscape

Population pressure continued to mount in the 1920s behind the city wall. When the Government announced that it would consider funding Improvement Schemes for Delhi in 1926, the DMC proposed demolishing the wall and acquiring the land behind it for improvement.[114] In his note entitled 'Improving the Delhi City,' made in March 1927, Mr Sohan Lal, the Secretary of the Municipal Committee, placed attention not only on dealing with congestion in the city via the Western Extension, but also on slum demolition.[115] He criticised the 'policy of drift' that the DMC and Government had adopted towards city extension and suggested the Delhi–Ajmeri Gate portion of the wall for demolition. The land could then be leased to poor people, preferably under a cooperative society, such that the people would come to own their land. The commentary by J.N.G. Johnson, the Deputy Commissioner, on the note stressed this proposal as amongst the most important in Lal's lengthy report.

Presumably encouraged, Lal wrote to Johnson in August 1927 requesting permission to not only demolish the wall but also to acquire the lands south of the wall towards the Circular Road.[116] While the DMC owned the land beneath the wall, the territory to the south was Nazul land that belonged to the Deputy Commissioner but was administered by the New Capital Committee. Johnson claimed that the Delhi Administration was already finding it difficult to get Government grants for the Western Extension and Daryaganj Nazul schemes, but the Committee said they would be willing to undertake the project themselves should they be given the land. The resolution of the Committee on 27 April stressed not only that such a scheme would benefit the economic development of the area and relieve congestion, but that it would also bring the two cities closer together, thus relieving many of the transport difficulties. It was this tension between the biopolitical and financial benefits versus the geographical effect on the distance, both mental and material, between the cities that so affected the views henceforth expressed of the project.

Given his constant pleading for slum clearance measures, it is not surprising that the scheme got the heartfelt endorsement of Dr Sethna. On 21 November 1927, Sethna stated that the pros outweighed the cons of the scheme and that while the wall was picturesque it had no historical interest and deprived the city behind of air and light. The park in no way served the function of a lung for the city and the new hospital being built near the parkland would in no way be affected by the town planned extension of the city south. He went on:

> Finally I would venture to point out the fallaciousness of not regarding the question of Delhi as a whole. Old Delhi and New are regarded as separate entities because separate bodies control them but really the question is one, namely the accommodation and well being of a large population which includes the Government of India for some five months in the year… Stone walls will not keep diseases away and so long as the really terrible present insanitary conditions prevail in the city, only a false sense of security can be felt with regard to the health of New Delhi, in spite of the modern sanitary conditions which prevail there.[117]

Sethna stressed that (due to the miscalculation of the accommodation required for the capital) thousands of clerks working in New Delhi had to live in the walled city. However, the Chief Engineer had argued on 22 September that he opposed the demolition of the wall due to its pleasing relief, its historical interest as a relic of the Old City, and also because of his fear of the expansion of the city 'without a break' beyond its existing limits. Many others raised such objections, to which Sohan Lal responded on 7 December 1927. He showed that fears about an interference with water supply were unfounded, that sentimentality had no place as a defence of the wall, and that the Government had had no problem in destroying other parts of the wall already. Finally, the argument that any Government grant would be an inappropriate subsidy was dismissed on the grounds that the congestion and inflation in the old city was *due* to the new city. The city wall, it was claimed, was the crux of the whole problem.

The DMC thus replied to the Chief Engineers concerns, following a meeting on 7 December 1927, that they still considered the scheme essential. Johnson informed the Chief Commissioner of their views in March 1928, yet Thompson only responded in May 1929. In reaching his decision, the Chief Commissioner admitted that while the Committee had some jurisdiction over the land, it had also been placed in the hands of the New Capital Committee and that it was they, in cooperation with the Nazul Department, who should propose schemes. As such, the proposal was rejected on three grounds. It was argued that the proposed hospital should not be near poor class housing and that the Government had already committed to other slum clearance schemes. Finally, the gardens that so beautified the border between the two cities was claimed to be of benefit to the slums should small openings be made in the wall.

The DMC responded 2 months later with a resolution of 10 July 1929 and a deputation to the Commissioner on 24 October.[118] The Western Extension was stressed as too remote for those living near Delhi Gate, while the health of the city was emphasised as a threat for the capital. The Chief

Commissioner responded with a press communiqué that stressed that the land to the south of the city *would* be used to relieve population pressure in Old Delhi. This was being done through building quarters for New Delhi clerks, thus reducing the number of Government workers, estimated at 15,000, who had to live in the old city. As a final insult, it was suggested that expanding the city might do nothing more than spread the slum south. This combination of lack of empathy with the plight of those behind the wall with a crass concern for the boundary between the two cities infiltrated much of the official discourse that would determine the fate of the wall over the following decade.

Despite Johnson's enthusiasm in 1927, the DAGS did not make it onto the list of Improvement Schemes that were financed by the central government in 1929.[119] There followed a period of inaction, while the congestion continued to rise behind the shield of the city wall. The scheme was given saddening impetus, however, during the monsoon of 1933. A portion of the city wall collapsed during the rains, killing two children and injuring a *tongawalla* (a rickshaw worker).[120] The question of demolition and improvement was raised in the Legislative Assembly on 1 September 1933, to which the Government claimed it was considering the matter.[121] The DMC forced the Deputy Commissioner, A.H. Layard, to order an assessment of the wall, which found it to be dangerous in parts.[122] On 26 October 1933, Layard wrote to the NDMC, which had inherited the land to the south of the city, suggesting that the wall was a liability and would have to be dismantled or repaired. The President of the NDMC responded with hostility to the idea of dismantling the wall, deploying all the symbolic potential of the capital area and of the old city as an unsanitary and dangerous place.[123] The letter of 6 December argued that it would be a sad fate for the New Capital if the wall was removed as the adjacent areas were bound to turn into 'insanitary dumping ground(s),' as had already happened near gaps and gateways in the wall. The President went on that:

> As the wall at present stands, it very appropriately defined the limits of the two cities, and keeps matters regarding the administration within bounds. If ever Government decided to demolish the wall, the NDMC would insist on an absolute unclimable fence being erected in its place, and erected before the wall was demolished.

Despite this trenchant criticism, the DMC demanded the demolition of the wall in a resolution of 31 May 1934.[124] Layard summarised the situation for the Chief Commissioner, showing that the NDMC had turned the *cordon sanitaire* land into a park, forcing the DMC to drop its more radical plans for southward expansion. Layard showed that this park was of no benefit as a lung for the city, that the wall was of no historical interest, and he went on to express his support for demolishing the wall.

While Layard had obviously dismissed the vigorous reaction by the NDMC to the perceived assault on their sovereign boundaries, H.S. Crosthwaite, the Chief Commissioner, had to mediate the views of the DMC with the more conservative opinions in New Delhi. On 27 August 1934, Crosthwaite wrote to the Secretary of Education, Health and Lands setting out the history of the DMC's requests and the NDMC's protests.[125] While sympathetic to the complaints from the capital that it would expose them to the view of the stables and slums behind, he was also conscious that the inhabitants could not enjoy the benefits of the park from behind an 8 m high wall. Thus, Crosthwaite sided with Thompson's recommendation of 1929 that holes be made in the wall for people, air and light to pass through.

In considering this situation, the Government acknowledged that there were three perspectives regarding the wall. The DMC wanted demolition, the Chief Commissioner wanted openings in the wall and the NDMC wanted its retention and repair. On 11 September 1934, R. Hutching from the Public Works Establishments Department, expressed his support for the NDMC's idea of a fence, which could also have openings in it, as it would '... serve the double purpose of protecting the park from damage and of providing a picturesque boundary between the Old and New Delhi Municipalities'. This combination of political security and aesthetic protection was gradually linked into the financial question of who should pay for the maintenance of the wall should it survive. The Punjab Municipal Act could be used to force the DMC to maintain the wall, while they would also be liable to pay for the demolition, if it took place, and to compensate the NDMC for any changes they would have to make to their park. Hutchins eventually came down in favour of openings in the wall. He suggested that one could hope no battlement was required to separate the inhabitants of the old city and the new. As such, he claimed that '[t]he conclusion is difficult to resist that the retention of the Wall is mainly justifiable on the ground that it will increase the amenities of the Park, both because it will form a pleasant back ground and will also hide the horrors which doubtless exist behind it' (see Figure 4.6).[126] As such, Hutching formed one of the first of an ongoing series of reactions within the Government of India against the political and aesthetic landscaping of the border to the detriment of the slum dwellers.

Viceroy Willingdon blocked this movement towards a partial renovation of the wall. He insisted in September 1934 that a slum clearance scheme would need to be planned and funded before the wall could be demolished, and stated that no openings in the wall should be made and any existing gaps should be repaired. In response, Hutching pointed out in January 1935 that the DMC did not want the wall so would be unlikely to repair it, and that the Government should pay for it themselves. On 18 March the Department of Industries and Labour recommended that the Government undertake the cost of repairing the wall and maintaining it on practical grounds until the

Figure 4.6   Slums inside the city wall

DMC had improved the slums behind. This was confirmed by the Legislative Department on 22 April 1935 when they showed that while the DMC *could* be forced to maintain the wall because it was in their jurisdiction, this also gave them the right to demolish it. As such, a grant-in-aid to the DMC was recommended, despite this having been turned down in the Improvement Schemes that were considered between 1926 and 1929.

This *volte-face* did not go unnoticed and on 23 April W. Christie, of the Education, Health and Lands Department, issued a withering attack on the policy decision. He claimed that:

> In the opinion of this Department it would seem wrong in principle to persuade the Municipality from carrying out their proposal by giving them a grant to cover the cost of the special repairs now necessary to the wall, as this would be tantamount to bribing the Municipality to refrain, for aesthetic and sentimental reasons and at the expense, and to the detriment, of the population of that portion of the City, from carrying out work which they otherwise consider necessary and advisable for sanitary reasons.[127]

Christie concluded that if the wall was of such historical importance it should be made a protected monument. This forced the Deputy Secretary of his own department to admit that the wall had no heritage justifications for being made a monument, and that the Viceroy might want to reconsider his opinion. While the Government was in general against the demolition of the wall without an improvement scheme in progress, it was admitted that the wall was dangerous and that the DMC could thus demolish it with every justification ahead of the monsoon. The Secretary of the Department, G.S. Bajpai, addressed the Viceroy on the issue and concluded on 8 May that it would be 'unwise to put the Municipality in a position to demolish the wall, at the same time abandoning the reported welcome move to clear up the slums behind the wall'. Bajpai also spoke to Christie and claimed, without reason, on 27 May that demolishing the wall would actually worsen the slums and pose a serious threat to the Irwin Hospital and New Delhi in general. As such, Bajpai made removal of the wall conditional on the slums being improved beforehand. While insisting that this was not bribery, he requested the Finance Department to pay for the maintenance of the wall and Rs 1.1 lakhs was provided in June 1935. However, as the Secretary of the DMC had written to the Chief Commissioner on 17 May, the entire DMC scheme was dependant on the removal of the wall and, as Crosthwaite had made clear in his original letter of August 1934, the Committee would not have the funds for the scheme for at least a few years. It was between these two strictures of the imperial ethos, the aesthetic and political fears of security combined with the financial 'bribes' and municipal short fallings, that the scheme stalled until the Trust took it up in 1937.

The DMC had made no effort to disguise its intentions regarding the scheme, with reports on its resolutions appearing in the *Hindustan Times* from March 1935. As such, the stalling of the scheme attracted coverage across the press. The *National Call*, a long-time critic of the Government, published an article on 18 July 1935 entitled 'The Tale of Two Cities, the Great Wall that Separates Old and New Delhi'. The article claimed that nowhere in India do two cities, 'one the last work on modern civic amenities and the other as primitive as the days of the cow and arrow civilisation' co-exist in such close contiguity. The violent contrast between these 'mileposts in the march of man's progress' was becoming obvious to those in the old city, who paid more money for far fewer amenities than those in New Delhi. This 'palpably invidious' distinction was said to be leading to growing discontent in the old city. The article condemned the Government for turning down the DMC's scheme, not because of sentimentality but because the removal of the wall would '... greatly detract from the beauty of New Delhi as the hideous slums which have hitherto been hidden by the lofty wall would be exposed to public view immediately on the borders of New Delhi.' This was claimed to be not just extremely selfish, but also without foundation as the point of the walls removal was to improve the slums.

The more moderate *Statesman* newspaper reported two days later on a Municipal Committee meeting that had re-confirmed the will to demolish the slums behind the Delhi–Ajmeri Gate wall. Mr Harish Chandra had stressed that the wall should go down so that the world would see the difference between the two Delhis. Similarly, Khan Bahadur Azizuddin claimed that 'The demolition of the wall would remove the *purdah* between Old and New Delhi and would provide purer air for the residents in that locality'. The use of term purdah is highly evocative. The term literally means 'veil' but was usually used to refer to the clothing worn by women throughout India to maintain their modesty in line with religious and cultural traditions. The term had been expanded to take on a geographical significance in terms of the confinement of women to the home, and even to certain sections of the house. Here, Old Delhi was positioned as not just the weak and feminine, but as the oppressed and the forcibly veiled. The financial and aesthetic machinations of the Government were thus posed as serving a political motive of concealment, one of which the local population was all too aware.

This criticism came at a crucial time. Pressure on the government had been growing since the 'Death Trap' article of January 1934 while the *Statesman* article of December 1934 also pointed out that the Government had dropped plans to demolish the wall and had turned the potential expansion area into a park. Why, they asked, was Delhi still without an Improvement Trust? The Delhi–Ajmeri Gate controversy had undoubtedly contributed to the pressure under which the Trust was formed. Hume had been serving as Deputy Commissioner since March 1935 and therefore took up his post at the head of the DMC as the press criticism of the scheme was at its height. Of the meeting on 18 July 1935 at which the Committee had confirmed they would pull down the wall, Hume wrote in his diary that he thought it was for the best.

The Committee issued a notification to acquire 8.8 acres of land behind the city wall in October 1935 and went about devising its own scheme to remove the wall, while the Government expended over Rs 1 lakh on maintaining it.[128] Because of this, Chief Commissioner Johnson's hope for the project was further diminished when he noted, on 13 June 1936, that the Committee had no plans to re-house the dispossessed slum dwellers, which he claimed would be necessary. In October, Johnson passed the scheme to Hume to see whether it would be worth recommending to the Government for funding, but Hume replied that the scheme was clearly a matter for the Trust and that it would be at the forefront of his proposals. When it became clear, following the Viceroy's note on the future administration of Delhi, that Hume would become the President of the NDMC, Chief Commissioner Jenkins thought it wise to inform him of some points of interest. He stressed that the New Delhi Committee had a great interest in the city wall removal scheme, and would insist upon a frontage of dignified buildings of uniform design as the boundary between the two municipalities. These two points, regarding the financial viability of re-housing the poor and the aesthetic landscaping of the boundary between the two cities, would continue to inform the Trust's attempts to push through the scheme.

Imperial finances and local resistance

The formation of the DIT in 1937 and the publication of its programme of schemes marked a defeat of sorts for the NDMC and certain people in the Government of India. Hume accepted the concept of the DAGS and vowed to demolish the wall, replacing it with a planned development of the slum areas. Hume worked to produce not only a frontage of unified buildings, but also came up against the financial strictures of the Government throughout his attempts to launch the scheme. Unlike the Western Extension (N7), the DAGS (T3) was not on Nazul land and thus the area had to be notified, acquired and constructed using either money raised from the Trust Account or using loans and grants from the Government. This effectively handed the central administration a vetting authority over the designs that they proceeded to use to guarantee that as little money as possible be spent on the scheme that could not guarantee a secure return.

Herein lay the origins of the clash between Hume and the Financial Department, perhaps the longest running and most passionate of his clashes within the state apparatus. Despite the somewhat abstract and mechanistic approach of the 1936 Report, Hume came to believe that it was the responsibility of the state to analyse housing at the human level and to admit that re-housing the poor could not be remunerative. At base, this represents a clash between an isolated state that sought to intervene only at moments of problematisation, and a welfarist theory of government that prioritised securing the processes of life for all, as a guarantee of a stable and secure population.

It was the issue of re-housing that most animated Hume throughout his work for the Trust. The Western Extension appeared more of a mess that had fallen on him to deal with, while the smaller schemes were worthy but, in the larger picture, of relatively little consequence. In the re-housing question, which was directly linked to the DAGS as the largest slum clearance scheme, Hume found an opportunity to launch a pioneering scheme in India, being the first place to apply the principles of caring for the poor which had been applied in the United Kingdom throughout the twentieth century.

In this commitment Hume was not entirely alone in Delhi, or without precedent. In Beadon's 1912 note on town planning the extension of Delhi, the question of re-housing the dispossessed was raised.[129] The labouring and menial classes that dwelled there were acknowledged as providing essential services for the city. Beadon thus recommended, if the residents were re-housed, that they would need not only residential accommodation, but also means of commuting to the city with convenience. Similarly, Craddock's response to de Montmorency's note on city expansion in 1912 had asserted that one of the general principles of town planning must be the provision of housing for the population displaced by town improvement.[130] Hume also had the support of the DMC in his attempt to create affordable housing for the poor, who had long been trying to clear the slums and move the people out of the city. Similarly, Major W.H. Crichton, who had experience of implementing the Housing Act (1935) in the United Kingdom, gave his support in the Health Report of 1937.[131] In it he stated that slums were growing faster than they were being cleared and that the founding policy of the Trust did not contemplate enough building of dwellings for those dispossessed by slum clearance. For Crichton, poor class housing was clearly the difficult question. It would be costly and there would be no guarantee that people would move there once houses were provided. However, this was the only viable way for the Trust, in his opinion, to actively *improve* and educate, and also to reduce the incidence of tuberculosis. As he put it, '... Government may lose money but this is impossible to avoid.' It was this British perspective on the economics of urban biopolitics that clashed so resolutely with the colonial Indian policy on town planning.

In the Report of 1936 Hume had quoted Sir Malcolm Hailey's view of the city wall as expressed during the Burn Bastion Road project.[132] Hailey had decried the wall as being damaging in terms of sanitation for the old city, in a bad state of repair, and both non-historic and non-aesthetic. Hume agreed and, in the programme of works for 1938–41, recommended the completion of a scheme inline with, but much larger than, the DMC's proposals.[133] The objects remained the same, namely the clearance of the unsanitary areas behind the city wall, the provision of light and air to a congested part of the city and the substitution of the wall for an elegant frontage facing New Delhi. However, the area required had increased from the 8.8 acres acquired to 39.9 acres, while 22.46 acres of wall, ditch and parkland would be taken from the Government.[134] The estimates were drawn up on the calculation of 200 people per acre, applying the norm Hume had devised in the 1936 Report.

Despite acknowledging it as an urgent measure, the Trust Administration Report for 1937–9 admitted that the scheme was still only in preparation.[135] A preliminary notification had been issued in 1938 for an area of 68 acres containing roughly 3,400 families, while the area had been surveyed and a layout prepared. However, it was claimed that the technical, architectural and administrative problems would still take time to mature. The reason for the scheme's slow progress cannot just be attributed to the financially crippling effects of the sewage works payments from the Trust Account, as the other extension works to the north of the city displayed similar patterns yet all entered construction in the early 1940s. The explanation for the deadlock lies not in the 280-year-old city wall, but in the as yet unbuilt houses for the poor and the demolition of the slums. With regard to these, Hume faced opposition from two directions, from above (the Government) and from below (popular protests).

Local resistance

In terms of the latter, some of the general worries about the Trust naturally came to settle on one of its largest projects, and that closest to most of the population of the city. One of the earliest meetings took place on 28 March 1938, as reported in the local newspaper *Watan*.[136] The meeting took place in one of the areas that had been notified under Trust law and sought to mobilise the notion of territorial rights against the Trust's legalistic powers. While the NDMC had earlier mobilised concerns over sovereignty through issues about borders at the level of the city, these protests mobilised a democratised and individualised notion of sovereignty in defence of the right to dwell where and how they liked. The protest was articulated in terms of *ancestral* properties being acquired and alienated. A further meeting on 29 March was attended by 500 people and was chaired by Municipal Commissioner Lala Ghasi Ram Lohia, who formulated objections to the acquisition of properties under the scheme. Local residents argued that their ancestral rights extended back over centuries and that their properties were more sacred to them than temples or mosques. They claimed to be content to go on living in the humble dwellings of their forefathers. It was also claimed that if the concern was with the appearance of the wall from New Delhi, then the Government should pay to improve the wall, rather than evict the city dwellers.

Hume commented in April 1938 that there was much agitation against the Trust, owing to inaccurate information on what was being done.[137] The Trust was bound by its own laws to give out lists of land that had been acquired, allowing agitators to stir up discontent in areas that had been targeted. The 4,000 acquisition notices that were being processed necessitated an 'objections subcommittee' while special attention was paid towards rumours that had been spreading in the press, against which Hume issued his own communiqué. Hume stressed that any notifications were for land needed for a public purpose and that each owner-occupier would have 60 days to send objections to the Trust and the right to be heard in person. A decision would only be made after hearing all these complaints, and the views of the DMC. The Chief Commissioner would then approve an acquisition, but even after this a Special Land Acquisition Collector would overlook the proceedings, against whose orders there was a right to appeal to a special tribunal. As Hume concluded, 'the rights of the people are fully safeguarded'. Indeed, the safeguards would lead to a substantial amount of the DAGS money being diverted to dealing with objections rather than acquiring land. This focus on the individual marked the shift in Hume's practice towards protecting the poor rather than driving through this showcase scheme, but this did not quell the popular protests against land acquisition.

During the six anti-DIT meetings that took place in late April 1938, that on the twenty-fourth registered the DAGS as the main source of concern, depicting it as a profit-making enterprise for the Municipality that would not re-house the poor.[138] The Trust issued another press note the following day stressing that notification was only the beginning of proceedings.[139] The note then tackled rumours that were being spread in an attempt to mobilise people living in affected areas. In terms of the poor, the note stressed that all objections would be heard and that re-housing arrangements would be made for those who required it. Against accusations that the scheme was only to the benefit of New Delhi, the note suggested that citizens in New Delhi were not interested in the squalor that lay behind the wall, which was true as long as it remained out of view. The idea of the Trust making a profit was dismissed as the DAGS was looking unlikely to break even. Finally, the idea that the scheme would benefit capitalists could not be dismissed, as land speculators were outside the Trust's remit, but it was stressed that mass hysteria and panic could only benefit private speculation.

The Indian National Congress capitalised on this tension, forming meetings near the slums to be demolished and claiming that Congress members would resign should the building go ahead. Twelve thousand members of the local community peacefully protested on 28 April against the development, which prompted a letter from the local Member of the Legislative Assembly to Jenkins. Mohammad Asaf Ali had been elected Delhi's representative and he had continued to push for urban reform in debating chamber. However, the protests united him with the Chief Commissioner, as he was regarded the 'original sinner' for having recommended the Trust originally and then championed the wall scheme. He had been targeted by panic-stricken residents and admitted that any land acquisition would arouse resistance. However, he stressed that previous schemes had ignored similar protests and had been successfully, and popularly, put through.

Asaf Ali then went on to outline his more fundamental misgivings about the approach of the trust that exposed the cultural geography of rights and the economic geography of business that underlay Hume's population geography of congestion. First, the middle and lower middle classes would be unwilling to leave their ancestral homes and *havelis* (mansions with interior courtyards). Yet, it was suggested that these 'moneyed' families should move out of the city because they could afford to, making space for the lower middle classes and poor within the city. These were people who needed to be close to their businesses and were often part of a shifting population of labourers, shop workers or artisans who would be drawn to areas associated with the *Karkhanadar* (manufacturing) class and who were attracted to the social life as much as the economic geography of the city. As such, Asaf Ali predicted that resistance would continue, and that much of it would be legitimate given the poverty of those to be evicted and their need to be near places of work. As a result of these types of agitation, Chief Commissioner Jenkins stated in May 1938 that the DAGS would be delayed while sites for re-housing were investigated.[140] The only way to successfully counter this resistance would be to have an effective re-housing policy and, after a trip to the United Kingdom, this was exactly what Hume sought to create.

Re-housing and welfare biopolitics

On 24 March 1938, at the beginning of the protests outlined above, Hume had admitted that there was still no clear policy on what would be done with the displaced population from the DAGS.[141] A week later Jenkins wrote to Hume that the question of the displaced population needed to be addressed, as they could not fend for themselves. Hume replied on 11 April that no firm policy could be decided upon until a census of the occupants had been carried out. However, for the poor occupants Hume stressed that his considered conviction was that the solution was *not* the provision of houses by the Trust. Rather, Hume favoured allotting land to displaced families at a concessional rent and constructing plinths on which housing could be built. Jenkins commented in a private note that this idea was disquieting and that Hume was still without a real idea with regard to re-housing.

The basis for Hume's future policy was laid in his 7-month leave to England between late April and late November 1938. He met town clerks at Leicester, Birmingham, Wolverhampton, Norwich and Coventry in an attempt to learn about town planning and the re-housing provisions of the Housing Act of 1935.[142] He would later say how crucial these experiences were for working out a Delhi scheme for re-housing the very poor.[143]

While Hume was away, Jenkins also came to favour the re-housing of the poor, stating in his reply to Asaf Ali on 28 April that re-housing was a statutory obligation, to all extent that is reasonable and necessary, in relation to a clearance scheme. Thus, the Trust had to consider the number to be re-housed, their location and the type of house. Jenkins wrote to Hume's temporary replacement and insisted that the Trust get on to the issue of re-housing straight away. Shortly before Hume's return, Crichton re-emphasised his support for re-housing. In a letter to the Superintending Engineer, the Chief Health Officer spoke of his experience with the Housing Act in the United Kingdom and stressed that such schemes could not pay for themselves. As such, 'The financial policy of the Trust should in my opinion be so arranged as to admit a loss in the improvement of a slum area like this but to recoup itself on the development areas such as the Western Extension.'

By the time of Hume's return he had accepted the necessity of re-housing and set about securing the finances necessary for it. He wrote to his parents on 12 March 1939 that he was fighting the Government over re-housing finances, but stated on 26 March that the Government was without vision and the DMC was wholly antagonistic. By 16 July Hume could write of a triumph with the Trust, having convinced the Government that decent housing conditions for the very poor was a public responsibility. The Government agreed to levy an Entertainments Tax in Delhi and to give the proceeds to subsidising the unremunerative part of re-housing in connection with slum clearance. Hume optimistically stated that 'It is the first move of its kind made by a government in British India and is entirely a step forward. We have guaranteed funds to build about 5000 homes.'[144]

In debating the nature of re-housing, Hume decided in September 1939 that his previous choice of allotting plots to evicted families had been unsatisfactory when used in the past.[145] Many had sold their plots at a profit and ended up filtering back into the city. Other houses that were built tended to be of an unsatisfactory standard, as they could not finance sanitary and durable housing. It was also acknowledged that housing the poor could not be an 'economic undertaking'. The decision had thus been made to provide houses for the poor using a public subsidy to cover the difference between the economic hire purchase instalment and the paying capacity of a family as assessed by the Trust. One, two and three bedroom houses were planned costing Rs 400, 560 and 700 to build respectively.[146] For a 20-year hire purchase scheme with maintenance and a concessionary ground rent the monthly instalment was estimated at between Rs 4 and 11, which the poor could mostly not pay more than Rs 3 per month from their average earnings of Rs 12. This led to what Hume termed a 'blaze of interest' in the Trust for having got the English re-housing policy accepted by the Government.[147] This involved the recognition that miserable housing is a public slur, that it was the Government's duty to do something about it, and that improvement was expensive and needed government funds.

The Superintending Engineer of Delhi Province was immediately ordered by the central government to see whether any economies could be imposed on the construction of the houses.[148] He reported on 2 December 1939 that the houses could not possibly be made any smaller, with thinner partitions, with thinner compound walls, or without their plastering. These designs were experimented with in the Arakashan (N8) and Hathikhana Schemes (T7) and showed in practice that the 137 re-housed families on average paid Rs 3 and 8 annas per month. Major Crichton's Health Report for 1939 hailed the Government's acceptance that slum dwellers would require funded housing as an advance of 'incalculable importance' of nationwide importance.[149] Hume began the year 1940 confident that he could address the 'real human problem' of slums and that he was going ahead with the first genuine plan for poor class re-housing ever undertaken by the state in India.[150]

Stalled improvement

Hume's optimism gradually faded as the Trust was forced to respond to local protests and the Government insisted on proof being provided that poor class re-housing could be provided on a large scale within budget. The adjusted Scheme was issued in April 1941 and had responded to criticisms from local persons and the DMC.[151] It was insisted that the area was a dire slum and that the scheme would remedy the major ills. The accusation that the proposed 'elegant façade' to face New Delhi would block as much light and air as the old wall was rejected because the new houses would be planned so as to allow better circulation and ventilation. Open spaces and amenities were provided and detailed plans were included to convince the DMC that the scheme had been thought through.

Despite the revised plans, by 6 September 1940, Hume was still requesting permission to start re-housing.[152] In a letter to the Chief Commissioner Askwith, he responded to the suggestion that the Trust's ground rent was not set high enough. He insisted that if the ground rent was set at the economic value of the land then the whole re-housing policy would have to be abandoned. People would not be able to pay the rent, and if the increase were to be met by an increased subsidy, this would not happen under the contemporary war conditions. Hume repeated his funding request for just 42 houses and forwarded the previous DIT resolution concerning the slums involved in the DAGS. It stated that 'The living conditions of the present group are of the worst description and in carrying out this small scheme the Board will be fulfilling to be best advantage one of its most sacred obligations.' Askwith forwarded Hume's request to the Education, Health and Lands Department insisting that a rise in ground rent would cut at the root of the Trust's re-housing policy.

On 10 January 1941, the Government granted this request, yet added that '... no further rehousing schemes will be considered until the working of the schemes not in force has been examined and a decision has been reached on the future policy to be adopted.' Hume reacted furiously, pitting his clash with the Government as a clear and moral opposition between economic and biopolitical concerns. On 23 February 1941, he wrote to his parents that the Trust was doing its best in wartime conditions, while facing the 'semi-civilised' attitudes of the Government:

> They have yet to learn that there are some things the success, nay, the urgency of which must be computed otherwise than in terms of rupees, annas and pies. The provision of conditions fit for human habitation is one of them. My particular game of bricks without straw is to produce slum clearance schemes which by hypothesis cannot show a profit in terms of rupees ... If I can do this the *bania*

[money making] instincts of the G of I [Government of India] grasp eagerly at it, if I can't they shake their heads sadly and say how sat it is that people must live in a mess, but surely it is none of their affair.

This marked Hume's last fracas with the Government over the DIT as, shortly afterwards, he left the Trust for another post. The debate over re-housing continued, however, and would delay the DAGS until after independence. This was despite the investigation showing that re-housing finances were generally sound, although adversely affected by the war economy. The new Chairman of the Trust, J.S. Hardiman, showed that these other schemes had been built to budget and had not proved unpopular with their targeted populations. The report from August 1941 showed that the other schemes, including the Andha Moghul (N6), Hathi Khana (T7) and Ara Kishan (N8) schemes, had only experienced problems with families wanting larger houses.[153] Askwith was convinced and wrote on 11 September 1941 that the Government should lift its embargo on re-housing schemes. This plea was unsuccessful and the DAGS was therefore stalled for another 2 years.

The following debate showed the Government to have almost totally reneged on its earlier promises to fund re-housing. On 26 January 1943, the Education, Health and Lands Department wrote to Askwith asking how the DAGS's anticipated deficit of Rs 1.74 lakhs was to be met.[154] Furthermore, the financial position of the Trust as a whole, in comparison to the anticipated position in the Hume Report, was drawn into question. Askwith responded in August that the first triennial report had returned four surplus and three deficit schemes, while the second programme of seven schemes anticipated a return surplus of Rs 2.4 lakhs. While this would probably be reduced, Askwith was confident that the DAGS deficit would be met by the surpluses of other schemes. The Chief Commissioner also had to fight off a proposed adjustment of the levelling between the Trust and Nazul accounts. The proposition would see the Nazul account benefit by Rs 3 lakhs at the expense of the Trust Account, which was already struggling to fund the re-housing. As such, Askwith urged a quick resolution of the DAGS standstill that had already been delayed for 2 years and was having such a bad effect on the Trust's reputation.

By December there had still been no breakthrough so Askwith decided to adopt a different method. In December 1941 Hardiman had accepted that the Trust was not going to win the Government over in terms of grants, so put in a request for all the profits from the Entertainments Tax.[155] The Government refused, but Askwith wrote to the DIT on 15 December 1943 encouraging them to press ahead with the re-housing schemes, despite the wartime rarity of materials and cost of construction. The Chief Commissioner stressed that many of the other Trust schemes could not progress without re-housing being provided and that even if these schemes could not be completed during the war, the housing could be made ready. Hardiman was encouraged to apply for a grant to cover the expanded construction costs, which could legitimately be requested from the Entertainment Tax fund. However, in May 1944 special conditions were brought to the Chief Commissioner's notice. The Public Works Branch had trebled their construction costs while the one-room tenements had proved unpopular in the re-housing schemes and would have to be dropped, thus increasing the cost of the schemes in total. These are the units that would have been occupied by the artisans and labourers Asaf Ali had drawn attention to, and who had proved reluctant to leave the city. The failure of the DAGS attracted growing criticism; including a *Hindustan Times* article of 1 October 1944 entitled 'Slummier and Slummier'. The article claimed that despite the efforts of Hume, Jenkins and Asaf Ali, the Government continued to block development of the south wall. It was claimed that the British had used the wall as a barrier to shut out of view the slums behind lest it offend the 'aesthetic sense' of those living in the capital.

Despite such criticisms, on 24 June 1945, W.T. Bryant, then Chairman of the Trust, was informed that finances for the DAGS had been suspended.[156] In addition, he noted that agitation was increasing amongst the owners of property that had been notified as under threat of acquisition since 1938. This had left them unable to sell their homes for a reasonable price, while further criticism had come from the press due to the considerable income accruing on Nazul lands, which few people realised could not be transferred to the Trust Account. Since abandoning the scheme was out of the question due to the effect it would have on the Trust's reputation, the only action was to postpone the scheme and revise upward the values of the properties under notification. The Land Development Officer of the Trust denounced this as 'legally unwise' as it increased compensation costs by 50 per cent from Rs 19 to 28 lakhs.[157] With such a burden the recommendation of postponement was passed and no further work took place on the scheme until after independence in August 1947.

The DAGS had undoubtedly been hit by the financial situation of the war economy. However, this alone cannot explain the dramatic failure of a scheme that had been desired and campaigned for since 1926. By 1943 the Roshanara City Extension was nearing completion and resulted in a profit of Rs 9.45 lakhs, the Northern City Extension was on the market by 1945, while even the Hathi Khana clearance scheme delivered a profit of Rs 1.04 lakhs.[158] The DAGS had not been part of a programme that would aestheticise the capital, on the contrary, it was deemed to pose a threat to the visual and physical barrier between the two cities. Nor had it been even potentially remunerative. As such, Hume had been fighting against the full force of the imperial ethos regarding landscaping and finance.

These events regarding the regulation of Delhi's population make tendencies visible that were both specific to Delhi, but also have wider resonances. There were continuities from the 1860s to the 1940s in the criticisms made of governmental inaction and the internal tensions this provoked. These ranged from Beadon's demand for 'ROOM,' to de Montmorency's insistence on Delhi's Western Extension, to Christie's assertion that the Government was bribing the Municipality into inaction, to Hume's vitriolic outbursts against his employers.

While emblematic of individual egos, temperaments and biographies, these statements were also situated in the structured and conditioning environment of colonial governmental finances. The Government of India had been shown, consistently, to govern at a distance. That is, its *laissez faire* approach withdrew the state to a point of minimal investment, and only at times of crisis was its conduct sufficiently problematised. In this case, problematisation resulted from the conflicting rationalities of the biopolitical and economic spheres, the former of which dragged in the state, the latter of which made it dig in its heels. In Europe there had been the realisation that the two domains were compatible: a healthy population led to a more robust economy and less social discontent. In India this realisation was forestalled by the difference of colonial governmentality.

This difference has been encapsulated in the abstract by the category of 'race'. On the ground this difference took up various spatial formations. The division between New and Old Delhi is an iconic representation of this difference between health and disease, order and disorder, boulevards and galis, white and brown. Not only was this division acknowledged but it was also defended, not just by Warrants of Precedence, but also by military cordon sanitaire, unclimable fences and elegant façades.

The refusal to adopt a welfarist regime of government beyond this difference in favour of a withdrawn, isolated state attracted as much criticism as did the constitutional or disciplinary acts of the Government of India. At the level of practical government the opposition and contestation of the regulation of population became clear. The image of the colonial government effortlessly interweaving the power and knowledge of census categories and population surveys into the lives of ordinary people thus becomes much complicated. The Western Extension and DAGS showed people taking up the categories of political society and articulating them in languages of petition, protest and democratised rights. This negotiation of difference enriches our understanding of colonial life, a richness that was also displayed in the face of the categorising landscape of New Delhi and the disciplinary surveillance across the two cities. Such resistance marks one of the many continuities across the landscapes explored in the last three chapters, as the conclusion will demonstrate.

Chapter Five

Conclusions: Within and Beyond the City

The previous chapters have analysed three landscapes of colonial ordering that responded, directly or indirectly, to problematisations provoked by the capital transfer of 1911. These three landscapes were unified by a predominant power relation and particular objects and subjects of ordering. However, as the introduction asserted, these power relations are intrinsically linked, and the targets of power were often the same. As such, there is a need to stress the complementary and co-constitutive nature of these landscapes. In addition, Delhi itself was as much relational as the landscapes it contained. The city not only formed a node within national and international networks, but also blurred temporal boundaries through its post-colonial effects. These relations will be hinted at to suggest a broader and more presentist genealogy of which this book may come to form a part.

In addition to being a genealogical thread within a history of Delhi's present, this research also has wider consequences. For Foucault scholars it has explored the practical intermeshing of sovereign, classificatory, disciplinary and biopolitical powers in a particular series of urban governmentalities. While each power type has been shown to have particular effects, the pre-eminence of governmental power has allowed an analytical, comparative approach to be adopted that teased out the regimes of government interlinking the three landscapes.

These interlinkages should also be of interest to post-colonial scholars in that they expose the 'difference' of colonial governmentality through its effect on people's lives and the spaces they occupied. This facilitates a critique of the effects of rule, rather than an attack on founding principles, philosophies or ideologies. Such a focus also leads into investigations of 'resistance' that go beyond Manichean divides and delve into acts that may be neither violent nor obvious. Drawing inspiration from post-colonial research, this focus on resistance has sought to overcome a shortcoming of the wider governmentality literature. Resistance has been shown as internal to government, or as entering into the apparatus as problematisations that provoke shifts in a regime that can spark further acts of resistance. Indeed, the case studies showed that most policies were not the product of an Enlightened, progressive ethos, but of a government responding to a threat to security. In this context it seems that Deleuze's (1988: 89; emphasis in the original) summary of Foucault is especially pertinent: '... the final word on power is that *resistance comes first*'.

Such an emphasis should also be of interest to colonial urban scholars and geographers, placing the agency of city dwellers at the heart of analysis. Focusing on the day-to-day activities of everyday lives, and the mundane practices of everyday administration, also helps overcome spatial Manichean divides between coloniser and colonised cities. This process is also aided by attention to the full variety of spatial formations through which colonial governmentalities operated, inserting an emphasis on 'place making' into governmentality approaches that can focus too much on spatial geometries and abstract processes of territorialisation. Such emphases have stressed formations ranging from infrastructures to imaginary geographies, technological spaces of calculation, to inhabited places of dwelling.

Finally, this book should be of interest to those performing archival research with theoretical questions and frameworks in mind. The genealogical methodology is self-consciously attuned to the archive; being '... grey, meticulous and patiently documentary ...' (Foucault, 1977: 139). It eschews a focus on monumental histories of key figures or moments, yet simultaneously seeks to avoid an antiquarian focus on continuity and stability. However, having extensively relied upon governmental archives, this book has risked a complicity with the government it seeks to critique, and an internal focus on a state that is supposedly governmentalised from outside. Yet, the heterogeneity and complexity of the colonial Indian archive gives a clear sense of the administration as a site of flux and indeterminism, although certainly also of incredible power. Requests, demands, innovations and anachronistic practices passed through and were issued forth, opening up as many avenues for enquiry as are closed down by the various silences and silencings of the archive. It is on these flows and connections that the rest of this chapter will reflect.

Interlinked Landscapes of Ordering

Practised connections

At the level of practice, in material context and embodied form, the three landscapes intermeshed and, at times, conflicted. This was very often a matter of not just *peopling*, in terms of populated and interacting space, but also of people involved in multiple projects who conjoined these landscapes.

The *residential* and *policing* landscapes of New Delhi have been shown to mutually constitute spatial ranks of prestige. The elite areas below Kingsway were intensely patrolled while the lower ranked, dual race, remainder of the city attracted less police manpower or finance, although still substantially more, per head, than the old city. The boundary between the two cities was protected through various plans while New Delhi could be condensed from the spacious residential landscape into an easily defendable Keep in times of need.

While the limited activities of the Delhi Improvement Trust (DIT) did not affect *policing* zones, the police were used for eviction and inspection purposes. The two landscapes were also linked by the interventions of administrators with their own perspectives on governmental action. J.N.G. Johnson and J.P. Thompson spent much of the 1930s alternating the role of Deputy Commissioner and Chief Commissioner. In 1928, Deputy Commissioner Johnson attempted to bring control of land use in New Delhi under the Chief Commissioner's control in order to ease congestion in the old city, and campaigned throughout the late 1930s in favour of the DIT. He also pleaded caution during the Civil Disobedience campaigns, arguing for 'moral' over 'physical' force. In contrast, Thompson was much more conservative in terms of allotting funds to the DIT and advocated moves in the disciplinary mechanism towards the use of tear gas, and the suitability of whipping small boys as punishment.[1]

Johnson's actions had displayed the connections between the *Trust* and *New Delhi*. The Trust struggled with its finances due to an administrative structure that had diverted funds to Nazul projects that benefited New Delhi, such as the malaria and sewage works. The remaining funds could not be invested in Old Delhi's southward expansion, as this project had been blocked due to aesthetic and epidemiological fears. The land itself had been used for a hospital, printing press and accommodation for the capital; the Western Extension was similarly appropriated. The projects were more deeply integrated in terms of cause and effect. The Trust had been established to remedy the congestion the capital transfer had brought about, but was financially handicapped from its very conception. The administration of the two Delhis highlights both the materially immanent potential for critique in the cities, but also what Dr Sethna referred to as '... the fallaciousness of not regarding the question of Delhi as a whole'.[2]

Analytical connections

The landscapes are united analytically in that they are all governmentalities, imbricating an element of biopower with longer standing relations of sovereign power. While not unified into one rigid regime of government, the analytical categories of the governmentality approach do allow comparisons to be drawn across the case studies.

In terms of *sovereign power*, the *residential landscape* of New Delhi represented the sovereign's right to order territory. As Foucault (1978b [2007]: 11 January) commented of La Maître's *La Métropolitée*, the capital city should not only be an ornament to the nation, but should also be a model of spatial distribution and political obedience and effectivity. Territory in New Delhi was divided and distributed by function (see Figure 2.1), and within function by rank so as to affirm the Warrant of Precedence that placed the sovereign King-Emperor's representative at the top. The sovereign authority to appoint offices was also used to guarantee that land rights would be administered by authorities concerned with protecting the capital not alleviating congestion in the walled city.

With regard to the *police*, sovereign power relations emerged most viscerally during anti-colonial and communal nationalist provoked periods of emergency. During these collapses of disciplinary power, the state of law was not withdrawn, but was redrawn to facilitate greater powers through the emergency ordinances and Criminal Procedure Codes of the 1930s, and greater violence by both the police and the military. Outside of these times and spaces of intense law, sovereignty functioned through 'moral effect', a spectacularly theatrical performance of sovereign power that took up the technology of urban discipline to an extent that was, in the words of Chief Commissioner Johnson, 'peculiarly prophylactic'.[3]

The DIT had to mould its activities around the fragmented Nazul sovereign land rights of the Government of India in Old Delhi. These lands had been violently appropriated from the previous Mughal sovereignty regime after the 1857 uprising and were used to orchestrate the improvement of the old city. The Trust was granted the Nazul land and yet had to pay Rs 200,000 per year for the privilege, positioning the central government as landlord and executor.

These articulations of sovereign power dovetailed with relations of *biopower* in the governmentalities that sought order in these landscapes. Since the different modalities of biopower cohered around administrative projects in Delhi they have been presented in separate chapters, but there were analytical continuities across these landscapes that mark them out as a unified regime of semi-autonomous governmental parts (see Table 1.1). The landscapes were all part of a colonial culture, explicated most clearly in New Delhi's residential landscape, which prioritised an *episteme* of classification and ranking. Whether routed through the power–knowledge relations of imperial or colonial urbanism or policing, spatial essentialism went hand in hand with an anthropological determinism that constructed race, class and gender into a matrix in which all could supposedly find their place.

This mindset produced *identity* assumptions in each grid of the matrix. White, upper class women in New Delhi were deemed useful within the home but incompetent out of it. Unable to identify the mimicking nationalist from the loyal subject, the population of Old Delhi was deemed to be wily and dangerous, never more so than when whipped up into a 'religious fervour' during communal processions and riots. In terms of improvement it was the absence of identity assumptions that marked the biopolitical governmentality. This was a by product of an episteme that overemphasised the classificatory urge, and calculated subjects as objects to be dispersed through the DITs schemes.

The identities of the subject population were objectified as such in all the landscapes, to some degree, through the distinctive yet interlinked modes of *visualisation*. In New Delhi the 1914 plan (Figure 2.1), though imperfectly realised by 1939 (Figure 2.2), clearly marked out a spatial hierarchy of precedence, much negotiated as this was on the ground. The threat of anti-governmental and communal riots forced a re-visualisation of the landscape in terms of the distribution of risk and surveillance (Figure 3.8), casting subjects as insurrectionists or fanatics. The DIT cast individuals as congested units on the intensity map of population that demanded levelling technologies to secure distribution and circulation (Figure 4.2). All three visualisations of these landscapes sought to abstract away from the individual onto a plain of uncontested knowledge and control that necessarily removed itself from individual need and agency.

This de-personalisation was also apparent in the colonial *techne*. As was fitting to a series of governmentalities bereft of pastoral or individualising care, the colonial apparatus emphasised technologies of governing from a distance rather than the deft conducting of conduct. Exceptions included the intensely scripted stage of New Delhi in which conduct as appropriate to rank was essential, or the achievement of discipline through theatrical 'moral effect', which was cheaper and involved less political risk. Elsewhere, the techne attempted to socially engineer order through zoned housing projects in New Delhi, surveillance that reported danger in a particular place and time, or infrastructural works that ineffectively attempted the dispersion of population intensity.

These forms of knowledge, identity, visibility and techne were infused with a colonial *ethos* that drew wealth, force and security to the top of the colonial hierarchy, and vociferously protected that position of privilege. In New Delhi this took the form of residential hierarchy, the police used zones of privilege and partition between the new and old cities, while the DIT worked under the rubric of de-congestion but funded the security of the capital.

The effects of these governmentalities enable a post-colonial critique that emerges from practice and spatial formation, rather than from a transcendental position or from the perspective of political philosophy. Following Young (2001), this critique is based on the problematisations that emerged at the time, whether explicitly anti-colonial or connected to mis-administration. It can emerge at any point within a government-ality, addressing sovereign, disciplinary of governmental power. As with the understandings of regimes of government, the critiques need not cohere into a devastating whole, but can display recurrent features relating to the effects of sovereign, or bio-, power.

With regard to the residential landscapes of New Delhi, the critique relates to the disparity between the government's sovereign powers and achieve-ments as a landlord and administrator. Though perfectly envisaged, the territory was not ordered in a manner that demonstrated the effectiveness of the sovereign. Due to this, the landscape had to be continually readjus-ted to protect the privilege of the elite. The intrinsic contradictions of these actions were highlighted through petitions and campaigns at the time, and provide us with a stark opportunity for a critique of colonial practice at the very heart of empire.

The sovereign violence of Delhi's police force strung together the fissures within its incomplete and ineffective surveillance network of anti-colonial and communal nationalist protestors. The violence was objected to at the time, in a general sense through Gandhi and the Indian National Congress's non-violent campaigns, but also through specifically local protests and spaces of withdrawal (see Legg, 2003, 2005b). Yet, beyond violence, the gov-ernment also tried to discipline the time–space rhythms of protest through demarcating the movement of communal processions. The spilling over of these processions was a rejection of colonial discipline, although in the name of a problematically violent communal politics than confuses any romanti-cised notion of Manichean virtue along the coloniser/colonised divide.

New Delhi's retention of sovereign land rights with regard to the DIT attracted due criticism. It was accused of failing, as a landlord, to govern its territory, but also of failing in its biopolitical task of knowing its people and its land. Hume unearthed the reason for this in his continual exposition of the tensions between the biopolitical and financial domains, while public commentators highlighted the government's concern regarding an imperial aesthetic for the museumified walled city rather than a modernising system of congestion relief.

While specific to each governmentality, these forms of critique unify as attacks upon the colonial ethos that circulated through each configuration of biopower and sovereign power, but retain their force through their spe-cificity. Yet, each act of critique must itself be situated in a wider genealogy. The tradition of petitioning was one that the colonial government brought with it, but which also stretched back to the culture of the Mughal court in Delhi. Forms of communal protest were used against the policing of public space, but predated the capital transfer and the modern classificatory pres-sures on previously more fluid forms of identity. Many of the critiques were unified by discourses and campaigns of an anti-colonial nationalism that had a logic and structure that was external to the government but recurrently penetrated it through externally originating problematisations. Accompanying the *Spaces of Colonialism* studied here were equally complex *Spaces of Nationalism* that merit study on their own grounds. Gandhi's protests drew not only on Indian history, but also on emerging forms of feminist and anti-colonial global protest. This was appropriate for a national protest against a globalised imperial power, in which the governmental practices of Delhi must now be relationally situated.

Beyond Colonial Delhi

Space

The approach to the spaces of colonialism here advocated has been inspired by a series of works which have also sought to undo the 'dual cities' hypothesis regarding colonial urbanism. Janet Abu-Lughod's (1965) assertion that the introduction of Western urban forms led to the segregation and isolation of native towns has left a great historiographical legacy. Further work by Abu-Lughod on 'urban apartheid' in Rabat (1980) has been complemented by work on Algiers (Çelik, 1997) and French policies in North Africa more generally (Rabinow, 1989; Prochaska, 1990; Wright, 1991). These works have outlined the processes of neglect and isolation that the authorities in New Delhi and the DIT eventually displayed with regard to Old Delhi. Yet, their emphasis on architectural space and the existence of two dual societies, one colonised and the other colonising, does not always reflect the intermingling of the two societies, nor the various types of spatial formations by which the colonisers attempted to order the native cities outside of which they dwelled.

Zeynep Çelik (1997: 5) has drawn attention to the 'complicated implications' hidden by the image of the dual city. While Algiers had the form of a dual city, it also played host to housing and urban policies that blurred that divide. Veena Oldenburg (1984) and Narayani Gupta (1981) have shown how the duality of north Indian colonial urban landscapes after the uprising of 1857 was violently transgressed not only to make physical space 'sanitary', but also to make the population 'safe and loyal'. Brenda Yeoh (1996) moved the object of study away from the socio-political context of architectural space to consider the ways by which Singapore was made visible and governable through physical, disciplinary and biopolitical spatial formations. Such studies suggest that the duality of colonial urbanism, which was erected and maintained in material space, was overcome via different forms of spatial ordering throughout the imperial world. They also provide additional evidence not for historiographical regicide, of ignoring the Western form in favour of native spaces, but of studying the interactions of the two and the interventions of the colonial government into seemingly segregated native spaces.

Beyond a comparative contextualisation, Delhi must also be situated within the many networks by which it was connected to cities within and beyond the subcontinent. As the capital city, *New Delhi* obviously functioned as a national and international showcase, putting forth a symbolic argument for the continuation of not just Indian colonialism but also British imperialism in the twentieth century. While Chapter 1 suggested that the city was read as much as a sepulchre or cenotaph of empire, its architectural influence has been much more widespread (Crinson, 2003). Yet, New Delhi was already an intensely international space, being architecturally inspired by the neo-classical Palladio traditions, the City Beautiful movement and the Garden Cities tradition within which Edwin Lutyens had trained. The city continued to function within international circuits of knowledge after its inauguration in 1931. The Chief Architect of the PWD suggested in February 1931 that all private designs in New Delhi must be by licensed architects, as he had noted was the practice in Singapore.[4] Similarly, when the problem with ribbon development along roads surrounding New Delhi emerged powers were adopted from the UK Ribbon Development Act (1935).[5]

In terms of *policing*, the policies in Delhi were placed under close scrutiny by the central Home Department which guaranteed the security of the capital. The police were also subject to memorandums sent to local governments and police forces throughout the country seeking continuity of policy with regard to sensitive issues such as the communal question. Yet, the Delhi police force was also subject to international influences. The riot force was directly modelled on the Shanghai Mob Street Platoon, while developments in the United States of America regarding the use of tear gas on civilian populations were monitored.

The DIT was created in part to challenge national assumptions that the capital city was falling into a state of decay. The calculations on which it was founded drew inspiration from former programmes of housing reform in Madras, while Hume had visited the Improvement Trust in Calcutta and been greatly impressed. The legal structure of the Trust itself was an amalgam of laws from Rangoon, Calcutta and the United Provinces. Hume clearly believed his designs would lead the way in India and was dismayed at the lack of commitment displayed by national representatives at a conference on industrial housing convened in January 1940.

Yet, the Trust was also international in inspiration. Hume drew on the techniques of the Colombo Ordinance (1919) but also explicitly set about applying the standards and technologies of the United Kingdom to Delhi following his British tour in 1938. In a *Hindustan Times* article of 27 November

1939, Hume openly envied not just the public conscience in the United Kingdom, but also the Addison Act (1919) and the Housing Act (1935) that had made urban reform possible. It was his attempt to overcome the colonial difference of investment in biopolitical government that would so frustrate his efforts in the late 1930s.

Time

A key facet of post-colonial studies has been the problematisation of the discontinuity of independence. Against notions of a temporal rupture have been posited transfers of power and continuities of life, government and exploitation. Yet, at first glance, the Indian example, and Delhi in particular, would seem to challenge this reading. With the partition of the subcontinent into India and Pakistan, east and west, and the subsequent migration, massive changes were wrought on the new states. After partition, 330,000 Muslims left Delhi while 500,000 non-Muslims arrived, overwhelming the urban infrastructure and setting the trend for continued in-migration that radically reconfigured the urban landscape (Pandey, 2001; Kaur, 2005). Broader governmental shifts also hinted at discontinuity, including the shift to development and democracy as key national ideologies.

Yet, the three landscapes outlined in this book proved resilient to the waves of people and ideas that washed over Delhi in the immediate post-independence period. Both the physical landscapes and the governmentalities of the colonial period exerted an influence over Delhi's post-colonial existence. There is not space here to discuss these 'postcolonial developmentalities' (Legg, 2006b) in full, but some suggestion of their continuities may be made.

New Delhi has been greatly transformed since 1947. Most of the clerks' quarters have been demolished and replaced with four-storey housing units, while grander commercial developments have continued to colonise the areas already penetrated by private enterprise in 1939. However, continuing the subzones of colonial privilege, a 'Lutyens bungalow' continues to stand as a status marker. The area to the south of Kingsway has been preserved as a heritage 'imperial zone' and is zealously defended (Dalrymple, 2004). The city is still protected by a *police* force that was further strengthened by Nehru and reached the zenith of their powers under Indira Gandhi's Emergency of 1975–77 (Tarlo, 2003).

The DIT was harshly criticised for its failings by an Enquiry Committee in 1950 and was replaced by the Delhi Development Authority. Although marked by a discursive shift from improvement to development, the new Authority maintained many of the calculations and assumptions of the DIT and continued to work towards a 'levelling of the intensity map' at a degree of abstraction that removed attention from what rival social reformers in Delhi at the time referred to as the 'human approach' (Mann, 2005b; Legg, 2006b, 199). None of these developments can be solely attributed to colonial influence (see Dupont et al., 2000), but they do encourage us to further investigate the effects of historical governmentalities on the present. This can be achieved through analysing the geographies of colonialism established in different forms of space, from the material to the social, and throughout the triangular relations of sovereign, disciplinary and governmental power.

Notes

CHAPTER ONE   IMPERIAL DELHI

1 *Punjab District Gazetteers, Volume V: Delhi District 1912* (Lahore: Civil and Military Gazette Press, 1913), 141–2.
2 *Second Report of the Delhi Town Planning Committee on the Choice of a Site for the New Imperial Capital* (India: HMSO, 1913), 59.
3 *The Daily Telegraph,* 19 February 1931.
4 *The Daily Herald,* 4 February 1931.
5 SASL/Shoosmith Papers/Box II.
6 *The Daily Telegraph,* 9 February 1931.
7 *The Hindustan Times,* 13 February 1931.
8 *The Observer,* 8 February 1931.
9 NMML/Legislative Assembly/1927/25 August/3374–5.
10 *The Times,* 11 February 1931.
11 *The Times,* 21 February 1931; *The Daily Mail,* 18 February 1931.

CHAPTER TWO   RESIDENTIAL AND RACIAL SEGREGATION

1 DA/DC/1917/47; DA/Home Education/1926/4(43).
2 *First Report of the Delhi Town Planning Committee on the Choice of a Site for the New Imperial Capital* (India: HMSO, 1913).
3 *Second Report of the Delhi Town Planning Committee on the Choice of a Site for the New Imperial Capital* (India: HMSO, 1913).
4 Cambridge University Library/Hardinge Papers/vol. 111/n. 32 GF De Montmorency to Sir James DuBoulay, 23 August 1912.
5 NA/PWD/Civil Works/Buildings A/May 1914/12–22.
6 DA/CC/Home/1927/145B.
7 NA/Home/Public/1929/482/29.
8 NA/Home/Public/1929/482/29.
9 NMML/Legislative Assembly/1931/9 March/1726–7.
10 NA/Home/Public/1934/1/40/34.

11 NA/Home/Public/1935/8/23/35.
12 NA/Public/Home/1928/8/11.
13 NA/PWD/Civil Works/Buildings A/September 1914/31.
14 DA/CC/Education/1933/4(79)B.
15 IORL/V/24/2983.
16 DA/CC/Finance/1938/102.
17 IORL/MSS EUR/C326.
18 SASL/Bayley (V.) papers. *One Woman's Raj*, memoir written in 1976.
19 DA/Confidential/1944/44C.
20 DA/Confidential/1944/44C.
21 DA/CC/Home/1928/105B.
22 NA/PWD/Civil Works/Buildings A/June 1917/1–4.
23 DA/CC/Home/1929/45B.
24 NMML/Legislative Assembly/1932/13 February/719–20.
25 *New Delhi Directory 1929–30* (Delhi: Model Press, 1930); *The Delhi Directory 1939–40* (Delhi: Publisher unknown, 1940).
26 IORL/EUR MSS/D724/8: A letter from A.P. Hume to his parents from 27 June 1936.
27 IORL/EUR MSS/D724/8: A letter from A.P. Hume to his parents from 17 April 1936.
28 IORL/EUR MSS/F183.
29 IORL/EUR MSS/D724/8: A letter from A.P. Hume to his parents from 13 September 1936.
30 Cambridge University Library/Hardinge Papers/vol. 113/62.
31 DA/CC/The New Capital Project Box 1/1912; Honourable Mr L. Porter's town planning file.
32 Cambridge University Library/Hardinge Papers/vol. 111/n. 102. Note by Cpt. G. Swinton on 19 December 1912.
33 Cambridge University Library/Hardinge Papers/vol. 112/n. 121a. Sir George Birdwood to Hardinge on 21 October 1913.
34 DA/DC/1925/10.
35 DA/CC/Foreign/1931/17B.
36 DA/CC/The New Capital Project Box 1/1912; Geoffrey de Montmorency, 18 March 1912.
37 Cambridge University Library/Hardinge Papers/vol. 110/n. 208. Hardinge to Sir Reginal Craddock, 22 April 1912.
38 NA/Public/Home/1928/8/11.
39 DA/CC/Home/1929/38B.
40 DA/CC/Home/1929/45B.
41 NA/Home/Public/1935/180/35.
42 Cambridge University Library/Hardinge Papers/vol. 111/n. 194. Hardinge to Sir Guy Fleetwood Wilson, 12 February 1913.
43 IORL/EUR MSS/E220/1A: Viceroy Hardinge to Sir Malcolm Hailey, 12 February 1913.
44 NA/PWD/Civil Works/Buildings A/May 1914/12–22.
45 NA/PWD/Civil Works/Buildings A/May 1914/12–22.
46 NA/PWD/Civil Works/Buildings A/September 1914/31.

47 NA/PWD/Civil Works/Buildings A/September 1915/7–8.

48 NA/PWD/Civil Works/Buildings A/June 1917/1–5.

49 IORL/L/PWD/6/1125.

50 IORL/V/24/4315: *Annual Progress Report for the New Capital Project at Delhi for the Year 1922–3* (Delhi: Superintendent of Government Printing, 1924).

51 NA/Home/Public/1925/583.

52 NA/Home/Public/1926/641/1/26.

53 NMML/Legislative Assembly/1928/19 March/1706–10.

54 DA/CC/Home/1929/45B.

55 Statement by the Rural Reconstruction League of India, 5 November 1932 (Churchill Papers, Churchill College, Cambridge, UK, CHAR/2/192/19–21).

56 DA/CC/Home/1929/45B.

57 NA/Home/Public/1929/1/26.

58 NMML/Legislative Assembly/1931/9 March/1726–7.

59 DA/DC/1916/17.

60 DA/DC/1918/29.

61 IORL/P/10347(1918, first half).

62 NA/Home/Public/1925/105.

63 DA/CC/Home/1927/38B.

64 DA/CC/Home/1927/145B.

65 Cambridge University Library/Hardinge Papers/vol. 112/n. 432. R.P. Russel, Secretary of the PWD, to Hardinge, 2 September 1915.

66 NA/Home/Public/1915 October/65 deposit.

67 Although after 1911 this term officially referred to people of mixed Indian and European decent (Blunt, 2005), it is being used here to refer to white clerks.

68 Cambridge University Library/Hardinge Papers/vol. 112/n. 432. R.P. Russel, Secretary of the PWD, to Hardinge, 2 September 1915.

69 NA/PWD/Civil Works/October 1915/65B.

70 NA/PWD/Civil Works/Buildings A/June 1917/1–4.

71 DA/CC/Local Self Government/Public Works Department/1938/1423.

72 DA/CC/Home/1929/45B.

73 NMML/Legislative Assembly/1921/21 March/1446.

74 NMML/Legislative Assembly/1923/22 March/3861.

75 NMML/Legislative Assembly/1927/14 September/4294.

76 DA/CC/Local Self Government/Public Health/1928/1373.

77 NMML/Legislative Assembly/1934/17 July/107.

78 DA/CC/Home/1929/45B.

79 NMML/Legislative Assembly/1921/19 September/415.

80 NA/Home/Public/1927/91/36.

81 NA/Home/Public/1929/1/26.

82 NMML/Legislative Assembly/1921/22 February/305–6.

83 NMML/Legislative Assembly/1925/7 September/796–7.

84 NMML/Legislative Assembly/1927/7 February/520.

85 NMML/Legislative Assembly/1931/924.

86 NMML/Legislative Assembly/1931/9 March/1727–8.

87 DA/CC/Home/1927/38B.

88 DA/CC/Home/1929/45B.

89 NA/Home/Public/1931/8/1/31.
90 DA/CC/Home/1927/145B.
91 NA/Home/Public/1931/8/1/31.
92 NA/Home/Public/1933/8/8.
93 NA/Home/Public/1939/31/23/39.
94 NMML/Legislative Assembly/1924/22 September/3891.
95 DA/CC/Home/1929/45B.
96 NMML/Legislative Assembly/1936/19 February/1124.
97 DA/DC/1926/15.
98 NMML/Legislative Assembly/1932/17 February/882–3.
99 NA/Education, Health and Lands/Health/1937/24–24/37-H.
100 DA/CC/Home/1929/45B.
101 NA/Education, Health and Lands/Local Self Government/1931/1–45A.
102 DA/CC/Education/1931/18B.
103 DA/CC/Education/1932/71B.
104 *New Delhi Development Committee 1939* (New Delhi: Government of India Press, 1941).
105 DA/CC/Local Self Government/1940/15(34).
106 NA/Home/Delhi/March 1913/15A.
107 NMML/Legislative Assembly/1937/1 October/2816.
108 DA/Confidential/1938/91C.
109 DA/DC/1939/97.
110 DA/Local Self Government/Public Works Department/1941/16(25).
111 *The Delhi Restriction of Uses of Land Act* (New Delhi: Government of India Press, 1942).
112 DA/DC/1939/97.
113 DA/CC/Miscellaneous/1943/9(61).
114 DA/DC/1926/15.
115 SASL/Stokes Papers/Box II, file 22, letters from MMW Yeatts to Lady Stokes, 21 March 1940. The French quotation is attributed to Marshall Pierre Bosquet, who made the comment on witnessing the charge of the Light Brigade in 1854. Many thanks to Mike Heffernan for this information.
116 IORL/MSS EUR/125/1.
117 NMML/Legislative Assembly/1943/7 November.
118 SASL/Bayley (V.) papers. *One Woman's Raj* memoir.

CHAPTER THREE   DISCIPLINING DELHI

1 NA/Home/Police/3/1927.
2 NA/Home/Police/3/1927.
3 *Report on the Administration of Delhi Province for 1927–8* (Calcutta: Government of India Central Publications Branch, 1929).
4 NA/Home/Police/3/1927.
5 *Report on the Administration of Delhi Province for 1923–24 to 1938–9* (Calcutta: Government of India Central Publications Branch, 1925–40); NA/Home Police/1942 75/3/42; DA/DC/1942/240.
6 DA/Home/Police/1931/11/XXII/31.

7   NA/Home/Police/1932/II/XIX/32.
8   Legislative Assembly Debate, 22 November 1940, 234.
9   NA/Home/Police/1943/74/10/43.
10 Census of India (1933). *Volume XVI: Delhi*. (Lahore: Civil and Military Gazette Press); Census of India (1942). *Volume XVI: Delhi*. (Simla: Government of India Press); NA Home (Police) 1943 74/10/43.
11 NA/Home/Police/1943/74/10/43.
12 DA/DC/1944/280.
13 Legislative Assembly Debates, 22 November 1940, 974–5.
14 Legislative Assembly Debates, 27 November 1940, 1099–100.
15 Legislative Assembly Debates, 27 October 1941, 513.
16 NA/Home/Police/1943/74/10/43; DA/1942/240.
17 NA/Home/Police/1943/74/10/43.
18 DA/Home/Confidential 1931/3(3)C.
19 NA/Home/Police/1930/74/VI/30.
20 NA/Home/Police/1930/74/VI/30.
21 NA/Home/Police/1930/74/VI/30.
22 NA/Home/Police/1931/54/IV/31.
23 NA/Home/Police/1943/74/10/43.
24 DA/Confidential/Military/1931/16B.
25 NA/Home/Political/1931/187.
26 NA/Home/Political/1938/21/46/38.
27 Cambridge University Library, Hardinge Papers/Kent Papers/Va28.
28 DA/Home/Confidential/1930/44B.
29 DA/Home/Confidential/1932/2(15)B.
30 NA/Home/Political/1938/21/8/40.
31 NA/Home/Political/1938/21/46/38.
32 NA/Home/Political/1942/3/34/42.
33 DA/Confidential/Military/1928/10B.
34 DA/Confidential/Military/1928/10B.
35 DA/Home/Confidential/1930/29B.
36 DA/DC/1942/381/(emphasis in the original).
37 DA/DC/1942/381.
38 DA/Home/Confidential/1942/128/42 C.
39 DA/DC/1942/381.
40 DA/DC/1942/381.
41 DA/Deputy Commissioner's files (henceforth DC)/1911/50.
42 DA/DC/1911/50/(emphasis added).
43 DA/Home/Confidential/1915/55.
44 DA/Home/Confidential/1915/55/(emphasis added).
45 J.A. Scott's Note in IORL/EUR/MSS/F161/75.
46 DA/DC/1920/59.
47 DA/Home/Confidential/1921/2B.
48 DA/Home/Confidential/1921/2B.
49 Legislative Assembly 8 March 1921, 717.
50 DA/Home/Confidential/1914/41B.
51 DA/Confidential/Military/1916/3B.

52 DA/Military/Confidential/1919/6B.
53 DA/Military/Confidential/1919/6B.
54 DA/Military/Confidential/1921/3B.
55 DA/Confidential/Military/1935/8B.
56 DA/Confidential/Military/1935/8B.
57 DA/Military/Confidential/1922/3.
58 DA/Home/1928/73B.
59 DA/Home/Confidential/1930/22B.
60 DA/Confidential/1936/1(17)B.
61 DA/Confidential/1936/1(17)B.
62 DA/Home/General/Confidential/1937/1(21)B.
63 DA/Confidential/1938/45-C.
64 DA/Military/Confidential/1921/11B.
65 DA/Military/Confidential/1921/11B.
66 DA/Military/Confidential/1922/4D.
67 DA/Military/Confidential/1932/11B.
68 DA/Home/Confidential/1920/33B.
69 DA/Home/Confidential/1920/33B.
70 *The Tribune*, 25 November 1921, India Office Records and Library L/R/5/202.
71 Three State Prisoners Regulations (Bengal Regulation III of 1918, Madras Regulation II of 1819 and Bombay Regulation XXV of 1827).
72 Bengal Criminal Law Amendment Act 1925.
73 Criminal Procedure Code (CrPC) Section 108.
74 Criminal Procedure Code (CrPC) Section 144.
75 Criminal Procedure Code (CrPC) Section 99a.
76 Part II of Indian Criminal Law Amendment Act 1908.
77 Indian Press Ordinance, 27 April 1930 (a revival of the Press Act 1910 that was repealed in 1922).
78 Prevention of Intimidation Ordinance, 30 May 1930.
79 Unlawful Instigation Ordinance, 30 May 1930.
80 Unauthorised News Sheets and Newspapers Ordinance, 2 July 1930.
81 Emergency Powers Ordinance, 4 January 1932.
82 DA/Military/Confidential/1930/2B.
83 DA/Military/Confidential/1930/2B.
84 DA/Confidential/1930/24C.
85 DA/Confidential/1930/24C/(emphasis in the original).
86 DA/Confidential/1930/24C.
87 DA/Confidential/Confidential/1930/12C.
88 DA/Home/Confidential/1930/55B.
89 DA/Home/Confidential/1930/28B.
90 DA/Home/Confidential/1930/28B.
91 DA/FR/5 May 1930.
92 DA/DC/1930/89.
93 DA/FR/21 July 1930.
94 *Hindustan Times*, 10 January 1932.
95 DA/Home/Confidential/1932/1(12)B; DA/Home/Political/1932/2(12).
96 DA/Home/Police/Confidential/1932/2(1).

 97 DA/Home/Confidential/1930/56B.
 98 DA/Home/1932/4(50)B.
 99 DA/Home/1932/4(50)B.
100 DA/DC/1942/381.
101 DA/DC/1942/381.
102 DA/DC/1942/381.
103 NA/Home/Political/1942/3/34/42.
104 DA/DC/1921/13.
105 Legislative Assembly Debate, 11 September 1924, 1781.
106 NA/Home/Political/1925/106/III.
107 DA/DC/1926/67; *Report on the Administration of Delhi Province for 1926–7* (Calcutta: Government of India Central Publications Branch, 1927).
108 DA/Home/Confidential/1927/18B.
109 DA/Military/1927/16B.
110 DC/1927/43.
111 DC/1927/45.
112 DA/DC/1933/55.
113 DA/Home/Confidential/1931/39/B.
114 DA/Home/Confidential/1935/1(7)B.
115 DA/Home/Confidential/1935/1(7)B.
116 DA/Home/Confidential/1935/1(7)B.
117 DA/DC/1934/45.
118 DA/DC/1943/55.
119 DA/DC/1935/45.
120 DA/DC/1934/34.
121 DA/DC/1942/134.
122 DA/DC/1935/46.
123 DA/Confidential/1938/1C; FR 2/2 November 1938.
124 DA/Confidential/1940/78/1/40C.
125 DA/Confidential/1940/78/1/40C.
126 NA/Home/Police/1942/75/3/42.
127 DA/Confidential/1940/78/1/40C.
128 NA/Home/Political/1939/22/15/39.
129 FR 2/2 August 1940.
130 DA/DC/1935/45.
131 DA/Confidential/1941/116/41C.
132 IORL/EUR/MSS/D724/10/8 December 1940.
133 DA/Confidential/1941/116/41C.
134 DA/Confidential/1942/56/1/42C.
135 DA/Confidential/1942/56/1/42C.
136 FR 2/2/November 1941.
137 IPI 9/P69.
138 FR 2/2 November 1941.
139 IORL/EUR/MSS/F161/197.
140 DA/Confidential/1946/113/46C.
141 DA/Confidential/1946/113/46C.
142 NA/Home/Political/1946/5/44.

143 NA/Home/Political/1946/5/44.
144 DA/Confidential/1946/113.

CHAPTER FOUR   BIOPOLITICS AND THE URBAN ENVIRONMENT

 1 IORL/V/27/230/12.
 2 NA/Home/Delhi/1912/May/52/A.
 3 NA/Home/Delhi/1912/May/52/A.
 4 SASL Hume Papers Box VII, *Report on the Relief of Congestion in Delhi: Volume I* (Simla: Government of India Press, 1936), 7 (hereafter referred to as *Relief of Congestion*).
 5 *Relief of Congestion*, 6.
 6 NA/Home/Delhi/1912/May/52/A.
 7 DA/CC/Education/1928/6(15)/B.
 8 DA/CC/Education/1930/6(26)B; *Health Bulletin: Public Health Report on the Delhi Province for the Year 1935* (Simla: GoI Press, 1937).
 9 *Report on the Administration of the Delhi Municipality for the Years 1929–30: Volume II, Annual Report of the Medical Officer of Health for 1929* (Delhi: Delhi Municipal Press, 1930), 2.
10 *Public Health Report on the Delhi Province for the Year 1930* (Calcutta: GoI Central Publications Branch, 1932).
11 DA/CC/Local Self Government/Local Bodies/1938/1335.
12 DA/CC/Education/1928/6(15)/B.
13 Legislative Assembly, 10 September 1928.
14 Legislative Assembly, 24 September 1929.
15 *Report on the Administration of the Delhi Municipality for the Years 1929–30, Volume II: Annual Report of the Medical Officer of Health for 1929* (Delhi: Delhi Municipal Press, 1930) 8.
16 NA/Education, Health and Lands/Local Self Government/1936/53-82/36-H.
17 DA/CC/Home/1930/29B.
18 DA/CC/Home/1930/29B.
19 NA/Education, Health and Lands/Local Self Government/1930/16B; DA/CC/Education/1935/51B.
20 DA/CC/Home/1934/144B.
21 DA/CC/Education/1929/4(105)B.
22 *Public Health Report on the Delhi Province for the Year 1930* (Calcutta: GoI Central Publications Branch, 1932).
23 *Public Health Report on the Delhi Province for the Year 1930* (Calcutta: GoI Central Publications Branch, 1932).
24 *Public Health Report: 1932* (New Delhi: Government of India Press, 1935).
25 DA/CC/Education/1934/6(1).
26 DA/CC/Education/1934/4(72).
27 DA/CC/Home/1930/29B.
28 *Health Bulletin: Public Health Report on the Delhi Province for the Year 1935* (Simla: Government of India Press, 1937).
29 Legislative Assembly, 12 September 1935.

30 All the quoted letters from Hume were written to his parents in England and are stored at IORL/EUR/MSS/D724. The years of the letters and their corresponding sub-files are as follows: 7 (1935); 8 (1936–7); 9 (1938–9); 10 (1940–1); 11 (1942–3). All dated references, unless stated, are from his correspondence.

31 Hume, 4 August 1935.

32 Hume, 6 September 1935.

33 Hume, 18–19 August 1935.

34 4 February 1936. Hume's diaries are also stored in the collection IORL/EUR/MSS/D724. The sub-references are as follows: 70 (1935–7); 72 (1939); 73 (1940–2).

35 Hume Diary, 6 February 1936.

36 Hume Diary, 7 February 1936.

37 *Report on the Relief of Congestion in Delhi: Volume I* (Simla: Government of India Press, 1936); henceforth referred to as *Relief of Congestion*.

38 *Relief of Congestion*, 5, 6.

39 *Relief of Congestion*, 15.

40 Taken from *Relief of Congestion: Volume II*.

41 *Relief of Congestion*, 27.

42 Hume, 5 May 1936.

43 Hume, 13 September 1936.

44 NA/ Education, Health and Lands: Health (Local Self Government)/1936/24-57/36-H.

45 'Chairman's memorandum to trustees, 19 March 1937' Box VII, Hume Papers, Centre for South Asian Studies, Cambridge.

46 NA/Education, Health and Lands/Local Self Government/1931/1-45A.

47 NA/Education, Health and Lands/Local Self Government/1931/1-45A.

48 NA/Education, Health and Lands/Health/Confidential/1937/23-49/37-H.

49 NA/Education, Health and Lands/Health/Confidential/1938/23-11/38-H.

50 Hume, 6 February 1937.

51 NA/Education, Health and Lands/Health: Confidential/1938/23-11/38-H.

52 Legislative Assembly, 26 January 1937.

53 DA/Confidential/Education/1937/12B.

54 DA/Confidential/Education/1937/12B.

55 DA/Confidential/Education/1937/12B.

56 Hume, 15 June 1937.

57 NA/Education, Health and Lands/Health/Confidential/1938/23-11/38-H.

58 NA/Education, Health and Lands/23-51/37-H.

59 NA/Education, Health and Lands/1937/23-51/37-H.

60 DA/CC/Local Self Government/Local Bodies/1938/794.

61 DA/CC/Local Self Government/Local Bodies/1938/794.

62 Hume, 11 August 1935.

63 *Delhi Improvement Trust: Application to Delhi of the United Provinces Town Improvement Act, 1919* (New Delhi: Government of India Press, 1937).

64 NA/Education, Health and Lands/Health/Confidential/1938/23-11/38-H.

65 *Delhi Municipality: Bye Laws, Rules and Directions* (Delhi: Municipal Press, 1937), 37.

66 *Delhi Improvement Trust: Application to Delhi of the United Provinces Town Improvement Act, 1919* (New Delhi: Government of India Press, 1937).

67 *Delhi Improvement Trust: Application to Delhi of the United Provinces Town Improvement Act, 1919* (New Delhi: Government of India Press, 1937).

68 Hume, 11 December 1938.

69 Hume, 26 March 1939.

70 Hume, 28 April 1940.

71 DA/Confidential/1938/98C.

72 DA/CC/Local Self Government/Local Bodies/1938/914.

73 DA/CC/Local Self Government/1939/1(117).

74 *Relief of Congestion*, 48.

75 *Administration Report of the Delhi Improvement Trust for Years 1937–9* (New Delhi: Delhi Improvement Trust, 1940).

76 *Administration Report of the Delhi Improvement Trust for Years 1937–9* (New Delhi: Delhi Improvement Trust, 1940).

77 *Administration Report of the Delhi Improvement Trust for the Years 1939–41* (New Delhi: Delhi Improvement Trust, 1942).

78 *Delhi Improvement Trust Three Year Programme, 1941–44* (New Delhi: Delhi Improvement Trust, 1941).

79 *Annual Public Health Report on Delhi Province for the Year 1937* (New Delhi: Government of India Press, 1939).

80 *Annual Public Health Report on Delhi Province for the Year 1938* (New Delhi: Government of India Press, 1939).

81 DA/CC/Local Self Government/Local Bodies/1938/1335.

82 DA/CC/Local Self Government/Local Bodies/1938/1335.

83 DA/CC/Local Self Government/1938/686.

84 DA/CC/Local Self Government/1938/686.

85 DA/CC/Local Self Government/1941/1(88).

86 Legislative Assembly, 27 March 1946.

87 *Relief of Congestion*, 7.

88 *Administration Report of the Delhi Improvement Trust for Years 1937–9* (New Delhi: Delhi Improvement Trust, 1940).

89 NA/Education, Health and Lands/Local Self Government/1924/28-9B.

90 DA/CC/Home/1930/29B.

91 DA/CC/Home/1930/29B.

92 Legislative Assembly, 10 September 1928.

93 *Report on the Administration of the Delhi Municipality for the Years 1929–30, Volume II: Annual Report of the Medical Officer of Health for 1929* (Delhi: Delhi Municipal Press, 1930).

94 *Report on the Administration of the Delhi Municipality for the Years 1930–31, Volume II: Annual Report of the Medical Officer of Health for 1930* (Delhi: Delhi Municipal Press, 1931).

95 *Relief of Congestion*, 10, 23.

96 Legislative Assembly, 24 March 1924.

97 DA/CC/Education/1934/4(72).

98 DA/CC/Education/1931/18B.
99 *Report of the New Delhi Development Committee 1939* (New Delhi: Government of India Press, 1941).
100 *Delhi Improvement Trust Three Year Programme, 1938–41* (New Delhi: Delhi Improvement Trust, 1939).
101 *Administration Report of the Delhi Improvement Trust for the Years 1939–41* (New Delhi: Delhi Improvement Trust, 1942).
102 DA/CC/Local Self Government/1938/686.
103 DA/CC/Local Self Government/Local Bodies/1938/1335.
104 DA/CC/Local Self Government/1938/686.
105 DA/CC/Local Self Government/1939/1(117).
106 *Annual Public Health Report for Delhi Province* (New Delhi: Government of India Press, 1941).
107 Legislative Assembly, 20 September 1939.
108 DA/CC/Local Self Government/Local Bodies/1940/1(44).
109 DA/CC/Local Self Government/Local Bodies/1940/2(106).
110 DA/CC/Local Self Government/Local Bodies/1940/1(85).
111 *Administration Report of the Delhi Improvement Trust for the Years 1939–41* (New Delhi: Delhi Improvement Trust, 1942).
112 DA/DC/1945/20.
113 Images from *T/3: Delhi Ajmeri Gate Slum Clearance and Development Scheme* (New Delhi: The Offset Art Press, 1941).
114 *Administration Report of the Delhi Improvement Trust for the Years 1939–41* (New Delhi: Delhi Improvement Trust, 1942).
115 DA/CC/Home/1930/29B.
116 DA/CC/Education/1928/4(100)B.
117 DA/CC/Education/1928/4(100)B.
118 DA/CC/Education/1929/4(105)B.
119 DA/CC/Home/1930/29B.
120 *Administration Report of the Delhi Improvement Trust for the Years 1939–41* (New Delhi: Delhi Improvement Trust, 1942).
121 Legislative Assembly, 1 September 1933.
122 DA/CC/Education/1934/4(187)B.
123 NA/Education, Health and Lands/Health/1934/24-25/34-H.
124 DA/CC/Education/1934/4(187)B.
125 NA/Education, Health and Lands/Health/1934/24-25/34-H.
126 *Administration Report of the Delhi Improvement Trust for the Years 1939–41* (New Delhi, Delhi Improvement Trust, 1942).
127 NA/Education, Health and Lands/Health/1934/24-25/34-H.
128 DA/CC/Education/1934/4(187)B.
129 NA/Home/Delhi/July 1912/11 Deposit.
130 NA/Home/Delhi/May 1912/52A.
131 DA/CC/Local Self Government/1938/499; *Annual Public Health Report on Delhi Province for the Year 1937* (New Delhi: Government of India Press, 1939).
132 *Relief of Congestion*, 41.

133 *Delhi Improvement Trust Three Year Programme, 1938–41* (New Delhi: Delhi Improvement Trust, 1938).

134 DA/CC/Local Self Government/1938/499.

135 *Administration Report of the Delhi Improvement Trust for Years 1937–9* (New Delhi: Delhi Improvement Trust, 1940).

136 DA/CC/Local Self Government/1938/499.

137 DA/Confidential/1938/98C.

138 NMML/Delhi CID/IX/98/1939/Chaman Lall Lohia.

139 DA/CC/Local Self Government/1938/499.

140 DA/CC/Local Self Government/Local Bodies/1938/914.

141 DA/CC/Local Self Government/1938/499.

142 British Library, Oriental and India Office, Private Collections, D724/9.

143 Hume, 19 November 1939.

144 Hume, 16 June 1938.

145 DA/CC/Local Self Government/Local Bodies/1939/1(103).

146 Image taken from The *Statesman*, 30 January 1940.

147 Hume, 19 November 1939.

148 DA/CC/Local Self Government/Local Bodies/1939/1(121).

149 *Annual Public Health Report on Delhi Province for the Year 1939* (New Delhi: Government of India Press, 1940).

150 Hume, 21 and 28 January 1940.

151 *T/3: Delhi Ajmeri Gate Slum Clearance and Development Scheme* (New Delhi, The Offset Art Press, 1941).

152 DA/CC/Local Self Government/Local Bodies/1940/1(100).

153 DA/CC/Local Self Government/Local Bodies/1941/1(29).

154 DA/CC/Local Self Government/1943/1(59).

155 DA/CC/Local Self Government/Local Bodies/1943/1(122).

156 DA/CC/Local Self Government/1943/1(59).

157 DA/CC/Local Self Government/1(52).

158 DA/CC/Local Self Government/1943/1(59).

CHAPTER FIVE   CONCLUSIONS

1 DA/CC/Home/1930/72B.

2 DA/CC/Education/1928/4(100)B.

3 DA/CC/Home/Confidential/1930/55B.

4 DA/CC/Education/1931/18B.

5 DA/Confidential/1938/91C.

Bibliography

Abu-Lughod, J. (1965) Tales of two cities: The origins of modern Cairo. *Comparative Studies in Society and History* 7, 429–57.

Abu-Lughod, J. (1980) *Rabat: Urban apartheid in Morocco.* Princeton, NJ: Princeton University Press.

Agamben, G. (1998) *Homo sacer: Sovereign power and bare life.* Stanford, CA: Stanford University Press.

Agamben, G. (2005) *State of exception.* Translated by K. Attel. London and Chicago: University of Chicago Press.

Allen, J. (2003) *Lost geographies of power.* Oxford: Blackwell.

Anderson, D.M. and Killingray, D. (1991) Consent, coercion and colonial control: Policing the empire, 1830–40. In D.M. Anderson and D. Killingray (eds) *Policing the empire: Government, authority and control, 1830–1940.* Manchester and New York: Manchester University Press, pp. 1–15.

Arnold, D. (1986) *Police power and colonial rule: Madras 1859–1947.* Delhi and Oxford: Oxford University Press.

Arnold, D. (1993) *Colonizing the body: State medicine and epidemic disease in nineteenth-century India.* Berkeley: University of California Press.

Ballhatchet, K. (1980) *Race, sex and class under the Raj: Imperial attitudes and policies and their critics, 1793–1905.* London: Weidenfeld and Nicolson.

Barret-Kriegel, B. (1992) Michel Foucault and the police state. In T. Armstrong (ed.) *Michel Foucault: Philosopher.* Hertfordshire: Harvester Wheatsheaf, pp. 192–8.

Barry, A., Osbourne, T. and Rose, N. (1996) *Foucault and political reason: Liberalism, neo-liberalism and rationalities of government.* London: University College London Press.

Bayley, E. (1980) *The golden calm: An English lady's life in Moghul Delhi.* Exeter: Webb & Bower.

Bayly, C. (1985) The pre-history of 'Communalism'? Religious conflict in India, 1700–1860. *Modern Asian Studies* **19**, 177–203.

Bayly, C. (1996) *Empire and information: Intelligence gathering and social communication in India, 1780–1870.* Cambridge: Cambridge University Press.

Behdad, A. (1994) *Belated travellers: Orientalism in the age of colonial dissolution*. Cork: Cork University Press.

Bhabha, H. (1994) *The location of culture*. London: Routledge.

Blake, S. (1991) *Shahjahanabad: The sovereign city in Mughal India, 1639–1739*. Cambridge: Cambridge University Press.

Blunt, A. (1999) Imperial geographies of home: British women in India, 1886–1925. *Transactions of the Institute of British Geographers* 24, 421–40.

Blunt, A. (2000) Embodying war: British women and domestic defilement in the Indian 'Mutiny' 1857–8. *Journal of Historical Geography* 26, 403–28.

Blunt, A. (2005) *Domicile and diaspora: Anglo-Indian women and the spatial politics of home*. Oxford: Blackwell.

Bopegamage, A. (1957) *Delhi: A study in sociology*. Bombay: University of Bombay.

Bose, S. and Jayal, A. (1997) *Modern South Asia: History, culture, political economy*. Delhi: Oxford University Press.

Brogden, M. (1987) The emergence of the police – the colonial dimension. *British Journal of Criminology* 27, 4–14.

Burchell, G., Gordon, C. and Miller, P. (1991) *The Foucault effect: Studies in governmentality*. London: Harvester Wheatsheaf.

Butler, J. (1997) *Psychic life of power: Theories in subjection*. Stanford, CA: Stanford University Press.

Çelik, Z. (1997) *Urban forms and colonial confrontations: Algiers under French rule*. Berkeley and London: University of California Press.

Çelik, Z. (1999) New approaches to the 'non-western' city. *Journal of the Society of Architectural Historians* 58, 374–88.

Chandavarkar, R. (1992) Plague, panic and epidemic politics in India, 1896–1914. In T. Ranger and P. Slack (eds) *Epidemics and ideas: Essays on the historical perception of pestilence*. Cambridge: Cambridge University Press, pp. 23–40.

Chatterjee, P. (1986) *Nationalist thought and the colonial world: A derivative discourse?* London: Zed for the United Nations University.

Chatterjee, P. (1993) *The nation and its fragments: Colonial and postcolonial histories*. Princeton, NJ: Princeton University Press.

Chatterjee, P. (1995) *Texts of power: Emerging disciplines in colonial Bengal*. Minneapolis: University of Minnesota Press.

Chatterjee, P. (2000) Two poets and death: On civil and political society in the non-Christian world. In T. Mitchell (ed.) *Questions of modernity*. London: University of Minnesota Press, pp. 40–7.

Chatterjee, P. (2001) On civil and political society in postcolonial democracies. In S. Kaviraj and S. Khilnani (eds) *Civil society: History and possibilities*. Cambridge: Cambridge University Press.

Chatterjee, P. (2004) *The politics of the governed: Reflections on popular politics in most of the world*. New York: Columbia University Press.

Chatterjee, S. and Kenny, J.T. (1999) Creating a new capital: Colonial discourse and the decolonization of Delhi. *Historical Geography* 27, 73–98.

Chopra, P. (ed.) (1976) *Delhi gazetteer*. Delhi: Delhi Administration.

Christensen, E.A. (1995) Government architecture and British imperialism: Patronage and imperial policy in London, Pretoria and New Delhi (1930–1931). Unpublished doctoral thesis, Northwest University, Illinois.

Cohn, B.S. (1983) Representing authority in Victorian India. In E. Hobsbawm and T. Ranger (eds) *The invention of tradition*. Cambridge: Cambridge University Press, pp. 165–210.

Cohn, B.S. (1996) *Colonialism and its forms of knowledge; the British in India*. Princeton, NJ: Princeton University Press.

Connolly, W. (2004) The complexity of sovereignty. In J. Edkins and V. Pin-Fat (eds) *Sovereign lives: Power in global politics*. New York and London: Routledge, pp. 23–40.

Cooper, M. (2004) Insecure times, tough decisions: The nomos of liberalism. *Alternatives* **29**, 515–33.

Craddock, S. (2000) *City of plagues: Disease, poverty, and deviance in San Francisco*. Minneapolis: University of Minnesota Press.

Crinson, M. (2003) *Modern architecture and the end of empire*. Aldershot and Burlington: Ashgate.

Curtis, B. (2002) Foucault on governmentality and population: The impossible discovery. *Canadian Journal of Sociology-Cahiers Canadiens De Sociologie* **27**, 505–33.

Dalrymple, W. (1993) *City of djinns: A year in Delhi*. London: Flamingo.

Dalrymple, W. (2004) The rubble of the Raj. *The Guardian*, November 13.

Dalrymple, W. (2006) *The last Mughal: The fall of a dynasty, Delhi, 1857*. Bloomsbury: London.

Darwin, J. (1999) What was the late colonial state. *Itinerario* **23**, 73–82.

Das, V. (1990) Introduction: Communities, riots, survivors – the South Asian experience. In V. Das (ed.) *Mirrors of violence: Communities, riots and survivors in South Asia*. Oxford: Oxford University Press, pp. 1–36.

Davies, P. (1985) *Splendours of the Raj: British architecture in India, 1660–1947*. London: John Murray.

Deacon, R. (2000) Theory as practice: Foucault's conception of problematization. *Telos* **118**, 127–42.

Dean, M. (1994) *Critical and effective histories: Foucault's methods and historical sociology*. London: Routledge.

Dean, M. (1996) Putting the technological into government. *History of the Human Sciences* **9**, 47–68.

Dean, M. (1998) Questions of method. In I. Velody and R. Williams (eds) *The politics of constructionism*. London, Thousand Oaks and New Delhi: Sage, pp. 182–99.

Dean, M. (1999) *Governmentality: Power and rule in modern society*. London: Sage.

Dean, M. (2002a) 'Demonic societies' liberalism, biopolitics, and sovereignty. In T. Blom Hansen and F. Stepputat (eds) *States of imagination: Ethnographic explorations of the postcolonial state*. Durham, NC, and London: Duke University Press, pp. 41–64.

Dean, M. (2002b) Liberal government and authoritarianism. *Economy and Society* **31**, 37–61.

Dean, M. (2002c) Powers of life and death beyond governmentality. *Cultural Values* **6**, 119–38.

Dean, M. and Henman, P. (2004) Governing society today: Editors' introduction. *Alternatives* **29**, 483–94.

Dean, M. and Hindess, B. (1998) *Governing Australia: Studies in contemporary rationalities of government.* Cambridge: Cambridge University Press.

de Benoist, A. (1999) What is sovereignty? *Telos* **116**, 99–118.

Deleuze, G. (1988) *Foucault.* London: Athlone.

Deleuze, G. and Guattari, F. (1987) *A thousand plateaus: Capitalism and schizophrenia.* Minneapolis: University of Minnesota Press.

Dillon, M. (1995) Sovereignty and governmentality: From the problematics of the 'New World Order' to the ethical problematics of the World Order. *Alternatives* **20**, 323–68.

Dillon, M. (2004) Correlating sovereign and biopower. In J. Edkins and V. Pin-Fat (eds) *Sovereign lives: Power in global politics.* New York and London: Routledge, pp. 41–60.

Donelly, M. (1992) On Foucault's uses of the notion 'biopower'. In T. Armstrong (ed.) *Michel Foucault: Philosopher.* Hertfordshire: Harvester Wheatsheaf, pp. 199–203.

Dossal, M. (1991) *Imperial design and Indian realities: The planning of Bombay city, 1845–1875.* Bombay: Oxford University Press.

Dreyfus, H. and Rabinow, P. (eds) (1982) *Michel Foucault: Beyond structuralism and hermeneutics.* Brighton: Harvester.

Driver, F. (1985) Power, space, and the body – a critical assessment of Foucault's discipline and punish. *Environment and Planning D: Society and Space* **3**, 425–46.

Driver, F. (1993) *Power and pauperism: The workhouse system, 1834–1884.* Cambridge: Cambridge University Press.

Duncan, J.S. (1990) *The city as text: The politics of landscape interpretation in the Kandyan kingdom.* Cambridge: Cambridge University Press.

Duncan, J.S. and Duncan, N. (1988) (Re)reading the landscape. *Environment and Planning D: Society and Space* **6**, 117–26.

Duncan, J.S. and Duncan, N. (2004) *Landscapes of privilege: Aesthetics and affluence in an American suburb.* London and New York: Routledge.

Duncan, N. and Legg, S. (2004) Social class. In J. Duncan, N. Johnson and R. Schein (eds) *A companion to cultural geography.* Oxford: Blackwell, pp. 250–64.

Dupont, D. and Pearce, F. (2001) Foucault contra Foucault: Rereading the 'Governmentality' papers. *Theoretical Criminology* **5**, 123–58.

Dupont, V., Tarlo, E. and Vidal, D. (2000) *Delhi: Urban space and human destinies.* Delhi: Manohar.

Edkins, J. and Pin-Fat, V. (2004) Introduction: Life, power, resistance. In J. Edkins and V. Pin-Fat (eds) *Sovereign lives: Power in global politics.* New York and London: Routledge, pp. 1–21.

Elden, S. (2001) *Mapping the present: Heidegger, Foucault and the project of spatial history.* London: Continuum.

Elden, S. (2005) Missing the point: Globalization, deterritorialization and the space of the world. *Transactions of the Institute of British Geographers* **30**, 8–19.

Elden, S. (2007) Strategies for waging peace: Foucault as *collaborateur.* In M. Dillon and A. Neal (eds) *Foucault: Politics, society and war.* London: Palgrave.

Evenson, N. (1989) *Indian metropolis: A view towards the West*. New Haven, CT, and London: Yale University Press.

Foucault, M. (1967) *Madness and civilization: A history of insanity in the Age of Reason*. London and Sydney: Tavistock.

Foucault, M. (1970) *The order of things: An archaeology of the human sciences*. London: Tavistock.

Foucault, M. (1972) *The archaeology of knowledge*. London: Tavistock.

Foucault, M. (1973) *The birth of the clinic: An archaeology of medical perception*. London: Tavistock.

Foucault, M. (1975–6 [2003]) *Society must be defended: Lectures at the Collège de France 1975–76*. London: Penguin.

Foucault, M. (1977) *Discipline and punish: The birth of the prison*. Harmondsworth: Penguin.

Foucault, M. (1978a [2001]) Governmentality. In J.D. Faubion (ed.) *Essential works of Foucault, 1954–1984: Power*, vol. 3. London: Penguin, pp. 201–22.

Foucault, M. (1978b [2007]) *Security, territory, population: Lectures at the Collège de France 1978*. Basingstoke and New York: Palgrave Macmillan.

Foucault, M. (1979a [2001]) 'Omnes et singulatum': Toward a critique of political reason. In J.D. Faubion (ed.) *Essential works of Foucault, 1954–1984: Power*. London: Penguin, pp. 298–325.

Foucault, M. (1979b) *The history of sexuality volume 1: An introduction*. London: Allen Lane.

Foucault, M. (1980) *Power/knowledge: Selected interviews and other writings, 1972–1977*. Brighton: Harvester.

Foucault, M. (1982 [2001]) The subject and power. In J.D. Faubion (ed.) *Essential works of Foucault, 1954–1984: Power*, vol. 3. London: Penguin, pp. 326–49.

Foucault, M. (1986a) *The history of sexuality volume 2: The use of pleasure*. London: Viking.

Foucault, M. (1986b) *The history of sexuality volume 3: Care of the self*. London: Allen Lane.

Fyfe, N. (1991) The police, space and society: The geography of policing. *Progress in Human Geography* 15, 249–67.

Gatrell, V. (1990) Crime, authority and the policeman state. In F. Thompson (ed.) *The Cambridge social history of Britain 1750–1950 vol. 3: Social agencies and institutions*. Cambridge: Cambridge University Press, pp. 244–50.

Gilmour, D. (2005) *The ruling caste: Imperial lives in the Victorian Raj*. London: John Murray.

Gooptu, N. (2001) *The politics of the urban poor in early twentieth-century India*. Cambridge: Cambridge University Press.

Gordon, C. (1991) Governmental rationalities. In G. Burchell, C. Gordon and P. Miller (eds) *The Foucault effect: Studies in governmentality*. London: Harvester Wheatsheaf, pp. 1–52.

Gordon, C. (2001) Introduction. In J.D. Faubion (ed.) *Essential works of Foucault, 1954–1984: Power*, vol. 3. London: Penguin, pp. i–xli.

Goswami, M. (2004) *Producing India: From colonial economy to national space*. Chicago and London: Chicago University Press.

Gregory, D. (1998) Power, knowledge and geography – the Hettner lecture in human geography. *Geographische Zeitschrift* **86**, 70–93.

Gregory, D. (2004) *The colonial present*. Oxford: Blackwell.

Gregory, D. (2007) Vanishing points: Law, violence and exception in the global war prison. In D. Gregory and A. Pred (eds) *Violent geographies: Fear, terror and political violence*. London and New York: Routledge, pp. 205–36.

Guha, R. (1997) *Dominance without hegemony: History and power in colonial India*. Cambridge, MA, and London: Harvard University Press.

Gupta, N. (1981) *Delhi between two empires, 1803–1931: Society, government and urban growth*. Delhi: Oxford University Press.

Gutting, G. (1989) *Michel Foucault's archaeology of scientific reason*. Cambridge: Cambridge University Press.

Hacking, I. (1986) Making up people. In T.C. Heller (ed.) *Reconstructing individualism: Autonomy, individuality, and the self in Western thought*. Stanford, CA: Stanford University Press, pp. 222–36.

Hall, P. (1988) *Cities of tomorrow: An intellectual history of urban planning and design in the twentieth century*. Oxford: Basil Blackwell.

Hannah, M. (1997) Space and the structuring of disciplinary power: An interpretative review. *Geografiska Annaler B* **79**, 171–80.

Hannah, M. (2000) *Governmentality and the mastery of territory in nineteenth-century America*. Cambridge: Cambridge University Press.

Hannah, M. (2007) Formations of 'Foucault' in Anglo-American geography: An archaeological sketch. In J. Crampton and S. Elden (eds) *Space, knowledge, and power: Foucault and geography*. Aldershot: Ashgate, pp. 83–106.

Hansen, T.B. (2005) Sovereigns beyond the state: On legality and authority in urban India. In T.B. Hansen and F. Stepputat (eds) *Sovereign bodies: Citizens, migrants, and states in the postcolonial world*. Princeton, NJ, and Oxford: Princeton University Press, pp. 169–91.

Hansen, T.B. and Stepputat, F. (2002) Introduction: States of imagination. In T. Blom Hansen and F. Stepputat (eds) *States of imagination: Ethnographic explorations of the postcolonial state*. Durham, NC, and London: Duke University Press, pp. 1–38.

Hansen, T.B. and Stepputat, F. (2005) Introduction. In T.B. Hansen and F. Stepputat (eds) *Sovereign bodies: Citizens, migrants, and states in the postcolonial world*. Princeton, NJ, and Oxford: Princeton University Press, pp. 1–36.

Hardy, A. (1993) *The epidemic streets: Infectious disease and the rise of preventative medicine, 1856–1900*. Oxford: Clarendon Press.

Harrison, M. and Worboys, R. (1997) A disease of civilisation: Tuberculosis in Britain, Africa and India, 1900–1939. In L. Marks and M. Worboys (eds) *Migrants, minorities and health: Historical and contemporary studies*. London: Routledge, pp. 93–124.

Hasan, M. (1995) *MA Ansari*. New Delhi: Publications Division, Ministry of Information.

Hazareesingh, S. (2001) Colonial modernism and the flawed paradigms of urban renewal: Uneven development in Bombay, 1900–25. *Urban History* **2**, 235–55.

Helliwell, C. and Hindess, B. (2002) The 'Empire of Uniformity' and the government of subject peoples. *Cultural Values* **6**, 139–52.

Hindess, B. (1996) *Discourses of power: From Hobbes to Foucault.* Oxford: Blackwell.

Hindess, B. (1997) Politics and governmentality. *Economy and Society* **26**, 257–72.

Hindess, B. (2001) The liberal government of unfreedom. *Alternatives* **26**, 93–111.

Hintze, A. (1997) *The Mughal Empire and its decline: An interpretation of the sources of social power.* Aldershot: Ashgate.

Hobsbawm, E. (1973 [1994]) *Revolutionaries.* London: Phoenix.

Hopkins, A. and Stamp, G. (2002) *Lutyens abroad: The work of Sir Edwin Lutyens outside the British Isles.* London: British School at Rome at The British Academy.

Hosagrahar, J. (1992) City as durbar: Theatre and power in imperial Delhi. In N. AlSayyad (ed.) *Forms of dominance; on the architecture and urbanism of the colonial enterprise.* Aldershot: Avebury, pp. 83–107.

Hosagrahar, J. (2005) *Indigenous modernity: Negotiating architecture and urbanism.* London and New York: Routledge.

Howell, P. (2000) A private Contagious Diseases Act: Prostitution and public space in Victorian Cambridge. *Journal of Historical Geography* **26**, 376–402.

Howell, P. (2004a) Race, space and the regulation of prostitution in colonial Hong Kong: Colonial discipline/imperial governmentality. *Urban History* **31**, 229–48.

Howell, P. (2004b) Sexuality, sovereignty and space: Law, government and the geography of prostitution in colonial Gibraltar. *Social History* **29**, 444–64.

Hunt, A. and Wickham, G. (1994) *Foucault and law: Towards a sociology of law as governance.* London: Pluto Press.

Hussain, N. (2003) *The jurisprudence of emergency: Colonialism and the rule of law.* Ann Arbor: University of Michigan Press.

Huxley, A. (1926) *Jesting Pilate: The diary of a journey.* London: Chatto and Windus.

Huxley, M. (2007) Geographies of governmentality. In J. Crampton and S. Elden (eds) *Space, knowledge, and power: Foucault and geography.* Aldershot: Ashgate, pp. 185–204.

Irving, R. (1981) *Indian summer: Lutyens, Baker, and imperial Delhi.* New Haven, CT, and London: Yale University Press.

Isin, E.F. (1998) Governing cities, governing ourselves. In E.F. Isin, T. Osborne and N. Rose (eds) *Governing cities: Liberalism, neoliberalism, advanced liberalism.* Toronto: York University, pp. 33–118.

Jain, A.K. (1990) *The making of a metropolis: Planning and growth of Delhi.* New Delhi: National Book Organisation.

Jones, C. (2000) Languages of plague in early modern France. In S. Sheard and H. Power (eds) *Body and city: Histories of urban public health.* Aldershot: Ashgate, pp. 41–9.

Jones, M. (2003) Tuberculosis, housing and the colonial state: Hong Kong, 1900–1950. *Modern Asian Studies* **37**, 653–82.

Joyce, P. (2003) *The rule of freedom: Liberalism and the modern city.* Verso: London.

Kalpagam, U. (2000) Colonial governmentality and the 'economy'. *Economy and Society* **29**, 418–38.

Kalpagam, U. (2002) Colonial governmentality and the public sphere in India. *Journal of Historical Sociology* **15**, 35–58.

Kaur, R. (2005) Planning urban chaos: State and refugees in post-partition Delhi. In E. Hust and M. Mann (eds) *Urbanization and governance in India*. New Delhi: Manohar, pp. 229–50.

Kaviraj, S. (1994) On the construction of colonial power: Structure, discourse, hegemony. In D. Engels and S. Marks (eds) *Contesting colonial hegemony: State and society in Africa and India*. London: British Academic Press, pp. 19–54.

King, A. (1976) *Colonial urban development: Culture, social power and environment*. London: Routledge and Kegan Paul.

King, A. (1992) Rethinking colonialism: An epilogue. In N. AlSayyad (ed.) *Forms of dominance; on the architecture and urbanism of the colonial enterprise*. Aldershot: Avebury, pp. 339–57.

Laclau, E. and Mouffe, C. (1985) *Hegemony and socialist strategy: Towards a radical democratic politics*. London: Verso.

Lal, K. (1999) *Revolutionary activities in Delhi*. Delhi: Agam Kala Prakashan.

Laurier, E. and Philo, C. (2004) Ethnoarchaeology and undefined investigations. *Environment and Planning A* **36**, 421–36.

Legg, S. (2003) Gendered politics and nationalised homes: Women and the anti-colonial struggle in Delhi, 1930–47. *Gender, Place and Culture* **10**, 7–27.

Legg, S. (2005a) Foucault's population geographies: Classifications, biopolitics and governmental spaces. *Population, Space and Place* **11**, 137–56.

Legg, S. (2005b) Sites of counter-memory: The refusal to forget and the nationalist struggle in colonial Delhi. *Historical Geography* **23**, 180–201.

Legg, S. (2006a) Governmentality, congestion and calculation in colonial Delhi. *Social and Cultural Geography* **7**, 709–29.

Legg, S. (2006b) Postcolonial developmentalities: From the Delhi Improvement Trust to the Delhi Development Authority. In S. Raju, M.S. Kumar and S. Corbridge (eds) *Colonial and postcolonial geographies of India*. London: Sage, pp. 182–204.

Legg, S. (2007) Beyond the European province: Foucault and postcolonialism. In J. Crampton and S. Elden (eds) *Space, knowledge, and power: Foucault and geography*. Aldershot: Ashgate, pp. 265–88.

Legg, S. (2008) Ambivalent improvements: Biography, biopolitics, and colonial Delhi. *Environment and Planning A* **40**.

Lemke, T. (2001) 'The birth of bio-politics': Michel Foucault's lecture at the Collège de France on neo-liberal governmentality. *Economy and Society* **30**, 190–207.

Levine, P. (2003) *Prostitution, race and politics: Policing venereal disease in the British Empire*. London: Routledge.

Low, G.C.-L. (1996) *White skins/black masks: Representation and colonialism*. London: Routledge.

Lowe, L. (1991) *Critical terrains: French and British orientalisms*. Ithaca, NY, and London: Cornell University Press.

Macleod, C. and Durrheim, K. (2002) Foucauldian feminism: The implications of governmentality. *Journal for the Theory of Social Behaviour* **32**, 41–60.

Major-Poetzl, P. (1983) *Michel Foucault's archaeology of western culture: Towards a new science of history*. Brighton: Harvester.

Manderson, L. (1996) *Sickness and the state: Health and illness in colonial Malaya*. Cambridge: Cambridge University Press.

Mann, M. (2005a) Turbulent Delhi: Religious strife, social tension and political conflicts, 1803–1857. *South Asia: Journal of South Asian Studies* 28, 5–34.

Mann, M. (2005b) Town-planning and urban resistance in the old city of Delhi, 1937–77. In E. Hust and M. Mann (eds) *Urbanization and governance in India.* New Delhi: Manohar, pp. 251–78.

Mathur, Y. (1979) *Quit India movement.* Delhi: Pragati.

Mbembe, A. (2001) *On the postcolony.* Berkeley and London: University of California Press.

Mbembe, A. (2003) Necropolitics. *Public Culture* 15, 11–40.

McNay, L. (1994) *Foucault: A critical introduction.* Cambridge: Polity Press.

Mehta, U.S. (1999) *Liberalism and empire: A study in nineteenth century British liberal thought.* Chicago: University of Chicago Press.

Merry, S. (2001) Spatial governmentality and the new urban social order: Controlling gender violence through law. *American Anthropologist* 103, 16–29.

Merry, S.E. (2000) *Colonizing Hawaii: The cultural power of law.* Princeton, NJ: Princeton University Press.

Metcalf, T. (1989) *An imperial vision: Indian architecture and Britain's Raj.* Berkeley: University of California Press.

Metcalf, T.R. (1994) *Ideologies of the Raj.* Cambridge: Cambridge University Press.

Mills, C. (2003) Contesting the political: Butler and Foucault on power and resistance. *The Journal of Political Philosophy* 11, 253–72.

Mitchell, T. (1988) *Colonising Egypt.* Cambridge: Cambridge University Press.

Mitchell, T. (1991) Preface to paperback edition. *Colonising Egypt.* Cambridge: Cambridge University Press, pp. ix–xv.

Mitra, A. (1970) *Delhi: Capital city.* New Delhi: Thompson Press.

Morris, J. (1983) *Stones of empire: The buildings of the Raj.* Oxford: Oxford University Press.

Nandy, A. (1983) *Intimate enemy; loss and recovery of self under colonialism.* Delhi: Oxford University Press.

Neal, A.W. (2004) Cutting off the king's head: Foucault's *Society must be defended* and the problem of sovereignty. *Alternatives* 29, 373–98.

Neocleous, M. (1996) *Administering civil society: Towards a theory of the state.* Basingstoke: Macmillan.

Nilsson, S.A. (1973) *The new capitals of India, Pakistan and Bangladesh.* Sweden: Studentlitteratur Lund.

Ogborn, M. (1993a) Law and discipline in nineteenth century English state formation: The Contagious Diseases Acts of 1864, 1866 and 1869. *Journal of Historical Sociology* 6, 28–55.

Ogborn, M. (1993b) Ordering the city: Surveillance, public space and the reform of urban policing in England, 1835–56. *Political Geography* 12, 505–21.

Ogborn, M. (1995) Discipline, government and law: Separate confinement in the prisons of England and Wales, 1830–1877. *Transactions of the Institute of British Geographers* 20, 295–311.

Ogborn, M. (1998) *Spaces of modernity: London's geographies, 1680–1780.* London and New York: Guilford.

Oldenburg, V.T. (1984) *The making of colonial Lucknow 1856–1877.* Princeton, NJ: Princeton University Press.

O'Malley, P. (1996) Indigenous governance. *Economy and Society* **25**, 310–26.

O'Malley, P., Weir, L. and Shearing, C. (1997) Governmentality, criticism, politics. *Economy and Society* **26**, 501–17.

Osborne, T. and Rose, N. (1999) Governing cities: Notes on the spatialisation of virtue. *Environment and Planning D: Society and Space* **17**, 737–60.

Pandey, G. (1990) *The construction of communalism in colonial North India*. Delhi: Oxford University Press.

Pandey, G. (2001) *Remembering partition: Violence, nationalism and history in India*. Cambridge: Cambridge University Press.

Parsons, C. (1926) *Notes on the administration of Delhi province*. Calcutta: Government of India Press.

Pasquino, P. (1991) Theatricum politicum: The genealogy of capital police and the state of prosperity. In C. Gordon, G. Burchell and P. Miller (eds) *The Foucault effect: Studies in governmentality, with two lectures by and an interview with Michel Foucault*. London: Harvester Wheatsheaf, pp. 105–18.

Pershad, M. (1921) *The history of the Delhi municipality 1863–1921*. Allahabad: The Pioneer Press.

Phillips, K. (2002) Spaces of invention: Dissension, freedom, and thought in Foucault. *Philosophy and Rhetoric* **35**, 328–44.

Philo, C. (1989) 'Enough to drive one mad': The organisation of space in nineteenth-century lunatic asylums. In J. Wolch and M. Dear (eds) *The power of geography: How territory shapes social life*. London: Unwin Hyman, pp. 258–90.

Philo, C. (1992) Foucault's geography. *Environment and Planning D: Society and Space* **10**, 137–61.

Philo, C. (2000) The birth of the clinic: An unknown work of medical geography. *Area* **32**, 11–19.

Philo, C. (2005) Sex, life, death, geography: Fragmentary remarks inspired by Foucault's population geographies. *Population, Space and Place* **11**, 325–33.

Pickett, B.L. (1996) Foucault and the politics of resistance. *Polity* **28**, 445–66.

Prakash, G. (1999) *Another reason: Science and the imagination of modern India*. Princeton, NJ, and Chichester: Princeton University Press.

Prakash, G. (2002) The colonial genealogy of society: Community and political modernity in India. In Joyce Patrick (ed.) *The social in question: New bearings in history and the social sciences*. London and New York: Routledge, pp. 81–96.

Prashad, V. (2001) The technology of sanitation in colonial Delhi. *Modern Asian Studies* **35**, 113–55.

Prochaska, D. (1990) *Making Algeria French: Colonialism in Bône, 1870–1920*. Cambridge: Cambridge University Press.

Rabinow, P. (1982) Ordonnance, discipline, regulation: Some reflections on urbanism. *Humanities in Society* **5**, 267–78.

Rabinow, P. (1983) Indian summer: Lutyens, Baker, and imperial Delhi. *Design Book Review* **1**, 20–1.

Rabinow, P. (1984) Biopower in the colonies (1). In C. Belisle and B. Schiele (eds) *Les savoirs dans les pratiques quotidiennes: recherches sur les représentations*. Paris: CNRS, pp. 196–208.

Rabinow, P. (1989) *French modern: Norms and forms of the social environment*. Cambridge, MA: MIT.

Rabinow, P. and Rose, N. (2003) Foucault today. In P. Rabinow and N. Rose (eds) *The essential Foucault: Selections from the essential works of Foucault, 1954–1984*. New York: New Press, pp. vii–xxxv.

Radley, H. (1928) The suppression of riots. *Journal of the United Service Institution of India* **58**, 28–37.

Ransom, J.S. (1997) *Foucault's discipline: The politics of subjectivity*. Durham, NC, and London: Duke University Press.

Raychaudhuri, S. (2001) Colonialism, indigenous elites and the transformation of cities in the non-Western world: Ahmedabad (Western Indian), 1890–1947. *Modern Asian Studies* **35**, 677–726.

Ridley, J. (2002) *The architect and his wife: A life of Edwin Lutyens*. London: Chatto & Windus.

Rose, N. (1996) Governing 'advanced' liberal democracies. In A. Barry, T. Osbourne and N. Rose (eds) *Foucault and political reason: Liberalism, neo-liberalism and rationalities of government*. London: University College London Press, pp. 37–64.

Rose, N. (1999) *Powers of freedom: Reframing political thought*. Cambridge: Cambridge University Press.

Said, E. (1978) *Orientalism: Western conceptions of the orient*. London: Routledge and Kegan Paul.

Said, E. (1986) Foucault and the imagination of power. In D.C. Hoy (ed.) *Foucault: A critical reader*. Oxford: Blackwell, pp. 149–55.

Said, E., Beezer, A. and Osbourne, P. (1993 [2004]) Orientalism and after. In G. Viswanathan (ed.) *Power, politics and culture: Interviews with Edward W. Said*. London: Bloomsbury, pp. 208–32.

Scott, D. (1995) Colonial governmentality. *Social Text* **43**, 191–220.

Scott, D. (1999) *Refashioning futures: Criticism after postcoloniality*. Princeton, NJ: Princeton University Press.

Sealey, N.E. (1982) *Planned cities in India*. London: Extramural Division of the School of Oriental and African Studies.

Sen, S. (2000) *Disciplining punishment: Colonialism and convict society in the Andaman Islands*. Delhi and Oxford: Oxford University Press.

Sengoopta, C. (2003) *Imprint of the Raj: How fingerprinting was born in colonial India*. London: Pan Macmillan.

Shields, R. (1996) A guide to urban representation and what to do about it: Alternative traditions of urban theory. In A. King (ed.) *Representing the city: Ethnicity, capital and culture in the 21st century metropolis*. London: Macmillan, pp. 227–52.

Shoosmith, A. (1931) The design of New Delhi. *Indian State Railway Magazine* **4**, 423–30.

Singh, S.R. (1979) *Urban planning in India: A case study of urban improvement trusts*. New Delhi: Ashish.

Smith, C. (2000) The sovereign state v Foucault: Law and disciplinary power. *Sociological Review* **48**, 283–306.

Sorkin, M. (1998) Chandigarh after Corbusier. *Architectural Record* **186**, 67–8.

Spear, P. (1973) *Twilight of the Mughuls; studies in late Mughul Delhi*. Karachi: Oxford University Press.

Spivak, G.C. (1988 [2000]) Can the subaltern speak? In D. Brydon (ed.) *Post-colonialism: Critical concepts in literary and cultural studies*. London: Routledge, pp. 1427–77.

Spivak, G.C., Landry, D. and McLean, G.M. (1996) *The Spivak reader/ selected works of Gayatri Chakravorty Spivak*. London and New York: Routledge.

Stamp, G. (1981) New Delhi. In E. Lutyens (ed.) *The work of the English architect Sir Edwin Lutyens (1869–1944)*. London: Arts Council of Great Britain, pp. 33–43.

Stein, B. (1998) *A history of India*. Blackwell: Oxford.

Steinberg, P.E. (2006) Calculating similitude and difference: John Seller and the construction of agents of empire. *Social and Cultural Geographies* 7, 687–707.

Stenson, K. (1998) Beyond histories of the present. *Economy and Society* 27, 333–52.

Stenson, K. (2005) Sovereignty, biopolitics and the local government of crime in Britain. *Theoretical Criminology* 9, 265–87.

Stoler, A.L. (1991) Carnal knowledge and imperial power: Gender, race, and morality in colonial Asia. In M. di Leonardo (ed.) *Gender at the crossroads of knowledge: Feminist anthropology in the postmodern era*. Berkeley: University of California Press, pp. 209–66.

Stoler, A.L. (1995) *Race and the education of desire: Foucault's history of sexuality and the colonial order of things*. Durham, NC, and London: Duke University Press.

Stoler, A.L. (2002) *Carnal knowledge and imperial power: Race and the intimate in colonial rule*. Berkeley and London: University of California Press.

Stoler, A.L. and Cooper, F. (1997) *Tensions of empire: Colonial cultures in a bourgeois world*. Berkeley and London: University of California Press.

Tadros, V. (1998) Between governance and discipline: The law and Michel Foucault. *Oxford Journal of Legal Studies* 75, 75–103.

Tambiah, S. (1996) *Levelling crowds: Ethnonationalist conflicts and collective violence in South Asia*. Berkeley and London: University of California Press.

Tarlo, E. (2003) *Unsettling memories: Narratives of the Emergency in Delhi*. London: Hurst.

Taylor, P. (2000) Sovereignty. In R. Johnston, D. Gregory, G. Pratt and M. Watts (eds) *Dictionary of human geography*, 4th edn. Oxford: Blackwell, pp. 766–7.

Thakore, M.P. (1962) Aspects of the urban geography of New Delhi, Unpublished doctoral thesis, University College, London.

Thomas, N. (1994) *Colonialism's culture: Anthropology, travel and government*. Oxford: Polity Press.

Thrift, N. (2007) Overcome by space: Reworking Foucault. In J. Crampton and S. Elden (eds) *Space, knowledge, and power: Foucault and geography*. Aldershot: Ashgate, pp. 53–8.

Tillotson, G.H.R. (1989) *The tradition of Indian architecture: Continuity, controversy and change since 1850*. New Haven, CT, and London: Yale University Press.

Times, T. (1930) *The Times Book of India; a reprint of a special Indian number of The Times, February 18th 1930*. London: Times.

Vale, L. (1992) *Architecture, power and national identity*. London: Yale University Press.

Valverde, M. (1996) 'Despotism' and ethical liberal governance. *Economy and Society* 25, 357–72.

Vaughan, M. (1991) *Curing their ills: Colonial power and African illness*. Cambridge: Polity Press.

Veer, P.v.d. (1994) *Religious nationalism: Hindus and Muslims in India*. Berkeley: University of California Press.

Volwahsen, A. (2002) *Imperial Delhi: The British capital of the Indian empire*. London and Munich: Prestel.

Walker, R.B.J. (2004) Conclusion: Sovereignties, exceptions, worlds. In J. Edkins and V. Pin-Fat (eds) *Sovereign lives: Power in global politics*. New York and London: Routledge, pp. 239–49.

Wolpert, S. (1977) *A new history of India*. New York: Oxford University Press.

Wright, G. (1991) *The politics of design in French colonial urbanism*. Chicago and London: University Press.

Yeoh, B.S.A. (1996) *Contesting space: Power relations and the urban built environment in colonial Singapore*. New York and Oxford: Oxford University Press.

Young, R.C. (2001) *Postcolonialism: An historical introduction*. Oxford: Blackwell.

Index

New in the **RGS-IBG** book series

Series editors: JOANNA BULLARD & KEVIN WARD
Loughborough University; University of Manchester

**Royal
Geographical
Society**
with IBG
Advancing geography
and geographical learning

Geomorphology of Upland Peat Erosion, Form and Landscape Change

MARTIN EVANS & JEFF WARBURTON *University of Manchester; Durham University*

Geomorphology of Upland Peat offers a detailed synthesis of existing literature on peat erosion, incorporating new research ideas and data from two leading experts in the field.

APRIL 2007 ~ 296 PAGES 1-4051-1507-6 *[ISBN13: 978-1-4051-1507-7]* HB £55.00

Spaces of Colonialism Discipline and Governmentality in Delhi, India's New Capital

STEPHEN LEGG *University of Nottingham*

Examines the residential, policed, and infrastructural landscapes of New and Old Delhi under British Rule.

MARCH 2007 ~ 308 PAGES
1-4051-5633-3 *[ISBN13: 978-1-4051-5633-2]* HB £55.00 / 1-4051-5632-5 *[ISBN13: 978-1-4051-5632-5]* PB £24.99

Publics and the City

KURT IVESON *University of Sydney*

Based on original empirical research conducted in Australia and the United Kingdom between 1996 and 2003, *Publics and the City* explores the ways in which the urban is used and produced in struggles to establish novel modes of public discourse and sociability.

MARCH 2007 ~ 248 PAGES
1-4051-2732-5 *[ISBN13: 978-1-4051-2732-5]* HB £55.00 / 1-4051-2730-9 *[ISBN13: 978-1-4051-2730-1]* PB £19.99

People/States/Territories The Political Geographies of British State Transformation

RHYS JONES *University of Wales*

People/States/Territories examines the role of state personnel in shaping, and being shaped by, state organizations and territories.

MARCH 2007 ~ 224 PAGES
1-4051-4033-X *[ISBN13: 978-1-4051-4033-1]* HB £55.00 / 1-4051-4034-8 *[ISBN13: 978-1-4051-4034-8]* PB £24.99

After The Three Italies Wealth, Inequality and Industrial Change

MICHAEL DUNFORD & LIDIA GRECO *University of Sussex; University of Bari*

After the Three Italies develops a new political economy approach to the analysis of comparative regional development and the territorial division of labour and exemplifies it through an up-to-date account of Italian industrial change and regional economic performance.

NOVEMBER 2005 ~ 376 PAGES
1-4051-2520-9 *[ISBN13: 978-1-4051-2520-8]* HB £55.00 / 1-4051-2521-7 *[ISBN13: 978-1-4051-2521-5]* PB £24.99

Putting Workfare in Place Local Labour Markets and the New Deal

PETER SUNLEY, RON MARTIN & CORRINE NATIVEL
University of Southampton; University of Cambridge; University of Glasgow

This book is the first comprehensive and authoritative analysis of the New Deal and examines how far the programme has succeeded in responding to the diversity of conditions in local labour markets across the UK.

DECEMBER 2005 ~ 256 PAGES
1-4051-0785-5 *[ISBN13: 978-1-4051-0785-3]* HB £50.00 / 1-4051-0784-7 *[ISBN13: 978-1-4051-0784-6]* PB £19.99

Blackwell
Publishing

For more information on these and other titles within the series, or to order online, please visit **www.blackwellpublishing.com**

Printed and bound by CPI Group (UK) Ltd, Croydon, CR0 4YY

27/10/2024

14580366-0004